Understanding
JOINTS

Understanding
JOINTS

A practical guide to their structure and function

Bernard Kingston

MA (Oxon), PGCE, DO
Senior Lecturer in Functional Anatomy
Clinic Tutor
British School of Osteopathy, London, UK
Registered Osteopath

STANLEY
THORNES

First published in 2000 by:
Stanley Thornes (Publishers) Ltd
Ellenborough House
Wellington Street
Cheltenham
Glos.
GL50 1YW
United Kingdom

00 01 02 03 / 10 9 8 7 6 5 4 3 2 1

A catalogue record for this book is available from the British Library

ISBN 0 7487 5399 0

Typeset by Northern Phototypesetting Co. Ltd., Bolton
Printed and bound in Great Britain
by Martins The Printers Ltd, Berwick upon Tweed

To the two dedicated and hard-working women in my family to whom I owe so much: my wife, Maxine, and my sister-in-law, Margarethe.

'Ultimately, there are no parts at all. What we call a part is merely a pattern in an inseparable web of relationships.'

Fritjof Capra

Contents

Acknowledgements

I am indebted to the following who have helped me in the writing or production of this book, either directly or indirectly: John and Dinah Badcock, for their help with the classical derivations and for their immense hospitality and enduring friendship; my brother Tony and his wife Margarethe, for rescuing me when the words refused to be processed; Simon Curtis, for the generous use of his library resources; Will Podmore and Lesley Fagan, librarians at the British School of Osteopathy; my niece Alex, for all the excitement of *ER*; Nicola and Susie, for being Nicola and Susie; the staff and students at the British School of Osteopathy, for their patience; May Corfield and Maxine, for administrative assistance; Tony Wayte, Helen Broadfield, and Chris Wortley, my editors; and all those at Stanley Thornes who have helped to produce this book.

Finally, special thanks to my family, and my family of friends who have not seen enough of me (though they may disagree); to my wife, Maxine, who has seen the home improvements falter; and to our puppies, Poppy and Jasper, who doggedly refused to let me forget their Kentish walks, and thereby kept me sane.

Bernard Kingston

Chapter 1

Introduction:
How to use this book

This book aims to encourage you, the reader, to develop awareness and understanding of how the joints of the body work, through a process of palpation, self-discovery and interactive learning. The text is designed to provide a clear introduction to the main joints, supported by study tasks which are frequently of a practical nature; clinical notes are also included at an introductory level. It is essentially a book for beginners, though it is assumed that most readers are (or have been) on foundation courses which involve familiarity with the basic concepts of anatomy and physiology.

I have deliberately refrained from recommending specific texts for further reading, for the simple reason that such lists are at best arbitrary and at worst quickly become out of date. I would prefer to encourage the reader to make their own enquiries with the help of their tutors and colleagues according to needs and interests; and also taking note of the expanding interactive resources now becoming available.

The most important resource available to you as you read this book is your own body. You may be surprised how much you can learn through palpation and examination of your own joints and muscles, and perhaps how much you will respect your body as a result. You should also try to learn through observation and palpation of your colleagues.

Since this book may be used at home, the term 'colleague' may be extended to include friends or family, but obviously frail or otherwise unsuitable individuals should not be asked to participate in practical tasks, for reasons of health and safety. The normal rules of professional conduct and common sense should apply.

Readers who have used the companion volume *Understanding Muscles* will be familiar with the fact that in many study tasks I encourage the reader to shade diagrams in colour, to highlight labels and annotations, and so on. I have continued that practice in this book and would encourage you to reinforce your learning in this manner.

I have tried, where possible, to avoid unnecessary duplication of the material in *Understanding Muscles*, though some duplication is unavoidable. For example, I thought it useful to list the muscles that produce joint movements next to a diagram illustrating the movements concerned.

It is also important to understand how the muscles that move joints produce those movements, but it would be tedious to constantly request in the study tasks that readers should 'check that you understand how the muscles produce the movement shown'. When this is required, I have reduced the advice given to the abbreviated message 'muscle check'.

Understanding Muscles will usually supply the details, but it is assumed that the reader will also be able to draw on library resources and possibly other interactive sources such as CD-ROMs.

In this context, it is hoped that the reader will have access to skeletal specimens. Even if plastic, these are extremely valuable in helping one to understand how joints work. Plastic models of the full-sized articulated vertebral column are widely available and are particularly important in helping one to understand movements of the spine. Frequent reference is made to these in the study tasks, where they are referred to as 'flexible plastic spines'.

I have tried to make the layout of the book logical. The reader may work through chapter by chapter or select individual chapters for study, according to need. Not all study tasks need to be attempted if the reader is confident about their knowledge in the subject; and some may be used for self-testing. Where possible I have used diagrams with numbered labelling, so that they may be used for self-testing purposes.

I have included 'clinical notes' in this book in order to link learning to clinical problems that may be of interest and relevance, mainly to students of courses in manual medicine. It is important to emphasise that the examples chosen are designed to enlighten the reader rather than to act in any way as a comprehensive guide. Students of orthopaedics and manual medicine will progress to more comprehensive texts and specialised courses in these subjects; but for the novice the linked examples may help to show how anatomical understanding provides a firm foundation for diagnosis and treatment strategies.

Although I have sought to adopt a learning-through-palpation approach in this book, the text is, nevertheless, often deliberately reductionist and mechanistic. Having trained as an osteopath, I am aware of the disadvantages of a reductionist approach; but I happen to believe that it is difficult for a novice to be truly holistic without first understanding something of the parts that produce the sum of the whole. This is particularly true of the vertebral column. I take full responsibility for this, but hope that the reader will be able to take the necessary overview and appreciate that the vertebral column and entire body do indeed function as an integrated whole.

Above all, I hope that you will find that this book makes learning by discovery enjoyable. Having taught first-year BSc students for a number of years, I have been constantly surprised by the dearth of practical guides to anatomical learning. I offer this book as an attempt to fill part of that gap.

Chapter 2

Anatomical terminology and movements

The terminology used in this book to describe muscles and their movements conforms to that used in standard anatomy texts.

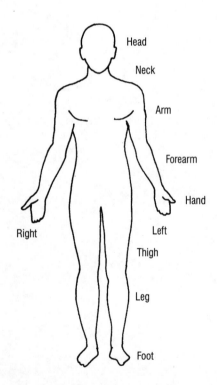

Figure 2.1A The anatomical position

Note: All anatomical descriptions relate to the anatomical position shown above. The conventional terms used to describe the limbs are also shown.

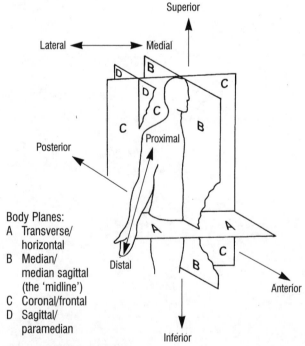

Body Planes:
A Transverse/ horizontal
B Median/ median sagittal (the 'midline')
C Coronal/frontal
D Sagittal/ paramedian

Figure 2.1B Body planes

Note: 'Medial' means closer to the median plane (B) and 'lateral' further from it. When applied to rotation movements, the terms 'internal' (medial) and 'external' (lateral) may be used.

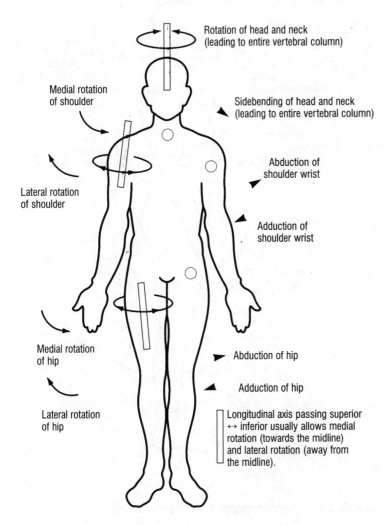

Rotation of head and neck
(leading to entire vertebral column)

Medial rotation
of shoulder

Sidebending of head and neck
(leading to entire vertebral column)

Abduction of
shoulder wrist

Lateral rotation
of shoulder

Adduction of
shoulder wrist

Medial rotation
of hip

Abduction of hip

Adduction of hip

Lateral rotation
of hip

Longitudinal axis passing superior
↔ inferior usually allows medial
rotation (towards the midline)
and lateral rotation (away from
the midline).

Figure 2.2A Anatomical movements: A/P and longitudinal axis

○ Antero-posterior (AP) axis passing front ↔ back through the body allows
abduction (movement away from the midline) and adduction (return to the
midline).

Note: Only a sample of the main axes are shown.

A number of specialized movements are also found which will be
described in the text as appropriate:

    – pronation/supination
    – inversion/eversion
    – dorsiflexion/plantar flexion
    – opposition
    – gliding
    – protraction/retraction
    – depression/elevation
    – circumduction

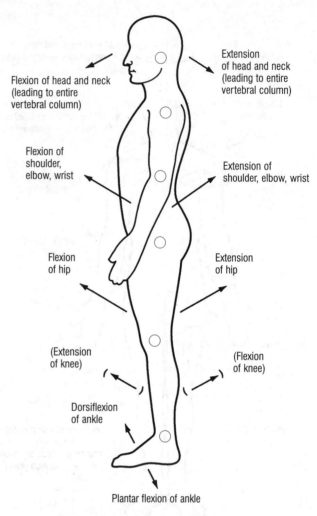

Figure 2.2B Anatomical movements: transverse axis

○ Transverse axis passing left ↔ right through the body allows flexion (approximating anterior surfaces) and extension (approximating posterior surfaces).

Note: Only a sample of the main transverse axes are shown.

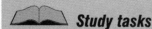 **Study tasks**

- Highlight the main details and features in each diagram.
- Shade the body planes in separate colours (p. 4).
- Shade in colour the open circles and lines denoting the axes of movement (pp. 5–6).

Chapter 3

Introduction to joints

The term 'joint' as used in this book will normally describe the zone where two or more articulating bones meet and produce movement. However, some textbooks commence the study of joints with an extensive discourse on the fuller meaning of the term, and it is important to appreciate that joints may also possess additional characteristics.

As well as acting as points of leverage and movement, joints may be regarded as zones where growth takes place and as sites that absorb and transmit forces. Since healthy function of joints depends on normal growth and usage, and since force transmission may become excessive, joints also tend to be focal sites of damage, and points for therapeutic intervention.

Note: Space permits only a brief introductory outline of the main characteristics of joints. Readers who require a more detailed account, or illustration of specific aspects, are referred to sources such as *Gray's Anatomy*, which is referred to in the text as Williams (1995).

A classification

Table 3.1 The main kinds of joints

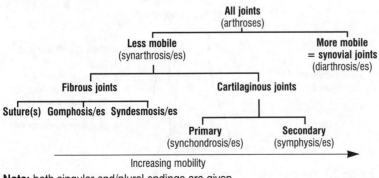

Note: both singular and/or plural endings are given.

 Study tasks

- Highlight the names of the main kinds of joints, using a different colour for the classical terminology (in brackets).
- Highlight the label 'Increasing mobility'.

The structure of joints

Fibrous joints

The articulating bones are united by fibrous connective tissue, and these joints are found where only a limited amount of movement is desirable.

Study tasks

- Highlight the details shown on the diagrams.
- Identify the examples on a skeleton and on yourself, as far as possible.
- Pause to consider the claims that the cranial bones move in order to facilitate the circulation of cerebrospinal fluid (Sutherland, 1990; Magoun, 1976).

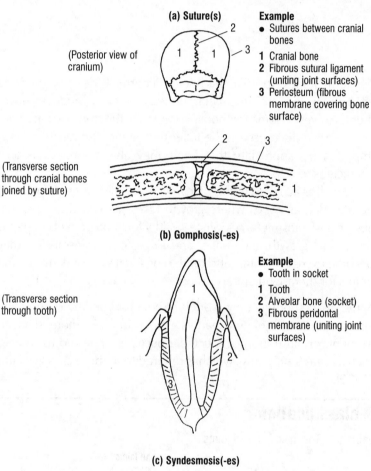

(a) Suture(s)

(Posterior view of cranium)

Example
- Sutures between cranial bones

1 Cranial bone
2 Fibrous sutural ligament (uniting joint surfaces)
3 Periosteum (fibrous membrane covering bone surface)

(Transverse section through cranial bones joined by suture)

(b) Gomphosis(-es)

(Transverse section through tooth)

Example
- Tooth in socket

1 Tooth
2 Alveolar bone (socket)
3 Fibrous peridontal membrane (uniting joint surfaces)

(c) Syndesmosis(-es)

(Coronal section of infe-rior tibiofibular joint)

Example
- Inferior tibiofibular joint

1 Tibia
2 Fibula
3 Fibrous interosseous liga-ment (uniting joint sur-faces)
4 Interosseous membrane

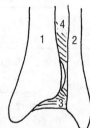

Figure 3.1 Fibrous joints: main types

Cartilaginous joints

The articulating bones are united by either hyaline cartilage, which is pearly, translucent and fairly elastic (primary cartilaginous joints), or fibrocartilage, which is fibrous and more resistant than hyaline cartilage. Joints that are united by fibrocartilage are referred to as secondary cartilaginous joints, or symphyses, and they may possess an internal cavity, or nucleus. Both types allow more movement than do fibrous joints.

(a) Primary cartilaginous joint (synchondrosis-es)

(Anterior view)

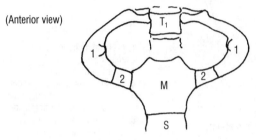

Example
- First sternocostal joint

1 First rib
2 Hyaline cartilage (uniting joint surfaces)
T₁ First thoracic vertebra
M Manubrium sterni
S Body of sternum

 Study tasks

- Highlight the details shown on the diagrams.
- Identify the examples on a skeleton and on yourself, as far as possible. The first sternocostal joint is palpable just inferior to the medial end of the clavicle (not shown in Figure 3.2(a)).

(b) Secondary cartilaginous joint (symphysis-es)

(Anterior view)

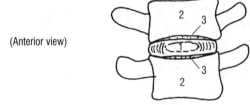

Example
- Intervertebral disc

1 Fibrocartilage disc with cavity (nucleus) (uniting joint surfaces)
2 Body of vertebra
3 Hyaline cartilage 'end plate' (covering articular bone surfaces)

Figure 3.2 Cartilaginous joints: main types

Synovial joints

These joints display the greatest degree of mobility and form the largest proportion of the joints studied in this book. Their unifying feature is that they possess a significantly large joint cavity, which is enclosed by a strong fibrous capsule, lined by synovial membrane (synovium) which exudes a specialised lubricant known as synovial fluid. 'Synovial' literally means 'with egg-like properties'. The fluid bathes the internal surfaces of the joint and the articular cartilage that covers the bone surfaces, acting as both lubricant and nutrient (articular cartilage has no direct blood or nerve supply). These features bestow a considerable degree of mobility, which exceeds that of fibrous and cartilaginous joints.

(a) Synovial joint: generalised structure

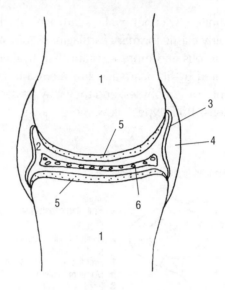

(b) Synovial bursa(-ae)

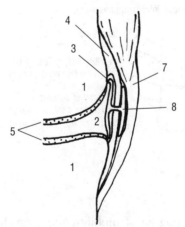

1 Articulating bone
2 Joint cavity
3 Synovial membrane
4 Fibrous capsule and ligament
5 Articular cartilage (usually hyaline, sometimes fibrocartilage)
6 Articular fibrocartilage disc (present in certain joints only)
7 Muscle and tendon
8 Synovial bursa

Figure 3.3 Synovial joints: generalised structure (diagrammatic sections)

The fibrous capsule is strengthened by strong bands of connective tissue, known as ligaments, which allow normal movements but tighten at the limits of joint movement to restrict injury. Unfortunately, they too may become damaged if movements are excessive. They may be found inside or outside the capsule, and if they provide support at the sides of a joint are usually called 'collateral' ligaments. All synovial joints receive additional support from surrounding muscles and their strong tendons. Tendons are composed of dense white connective tissue with half the tensile strength of steel and may play an important role in helping to protect joints.

In order to prevent friction between the sliding surfaces of ligaments, tendons and capsule, the synovial membrane may emerge from gaps in the capsule to form a 'bursa' (plural: bursae). This is a pouch or sac of synovial membrane which forms an interface between vulnerable sliding surfaces but may itself become vulnerable to injury.

If the articulating joint surfaces are not particularly congruent (do not show good fit) a fibrocartilage disc may be present within the joint cavity, which will tend to improve congruency and shock absorption (Figure 3.3(a)).

If only two articulating surfaces are present, the joint is known as a 'simple' synovial joint; if more than two surfaces are involved, it may be referred to as a 'compound' joint.

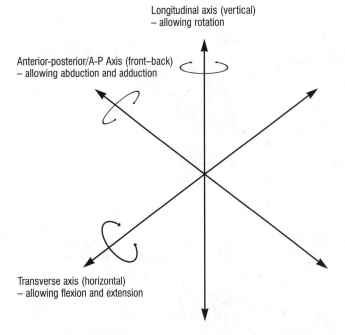

Figure 3.4 Axes of movement (showing three degrees of freedom)

The greater range of movement of synovial joints is a result of the several different types of joint surfaces, which enable movement to take place through either one axis (uni-axial), two axes (bi-axial), or three (multi-axial/polyaxial). In this particular classification, there are three possible 'degrees of freedom' of movement available

---

Clinical Note

Injury to the internal and external structures of synovial joints is common, as a result of the considerable mechanical forces that pass through these joints. Inflammation of various structures may be identified by the suffix '-itis'. Thus, inflammation of the synovial membrane is 'synovitis'; inflammation of a bursa is 'bursitis'. Inflammation of an entire joint is referred to as 'arthritis', which is a rather general term, and may be the result of degenerative changes or due to systemic disease.

 **Study tasks**

- Highlight the details shown in Figure 3.3.
- Shade the features of a synovial joint in separate colours as appropriate.
- Consider the factors that aid the stability of synovial joints.

through a transverse axis, an antero-posterior axis and a longitudinal axis. It does not, however, allow for the multi-directional movements that take place at plane synovial joints (Figure 3.5(g)), which may be described by the terms 'slide', 'glide' or 'translation'.

(a) Hinge
– allowing flexion/extension only, through transverse axis

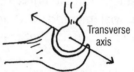

Transverse axis

Examples

- Humero-ulnar joint (elbow)

(b) Pivot
– allowing rotation only, through longitudinal axis

Longitudinal axis

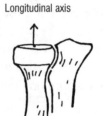

- Proximal radio-ulnar joint (elbow)

(c) Condylar/bicondylar
– allowing flexion/extension through transverse axis with limited rotation through longitudinal axis

Longitudinal axis

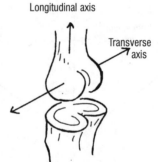

Transverse axis

- Tibiofemoral joint (knee)

(d) Ellipsoid
– allowing flexion/extension through transverse axis and abduction/adduction through A-P axis

A-P axis

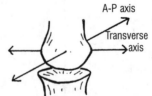

Transverse axis

- Metacarpophalangeal joint (hand)

(e) Saddle (sellar)
– as in ellipsoid, but more open surfaces may allow some rotation through longitudinal axis; the reciprocal surfaces are concavoconvex like a saddle

A-P axis

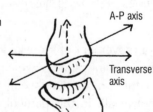

Transverse axis

- 1st carpometacarpal joint (thumb)

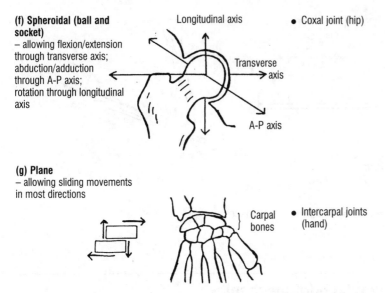

(f) Spheroidal (ball and socket)
– allowing flexion/extension through transverse axis; abduction/adduction through A-P axis; rotation through longitudinal axis

Longitudinal axis

Transverse axis

A-P axis

• Coxal joint (hip)

(g) Plane
– allowing sliding movements in most directions

Carpal bones

• Intercarpal joints (hand)

Figure 3.5 Synovial joints: main types

Levers

A lever is a rigid rod that rests on and acts over a relatively fixed point known as a fulcrum. In the human body the bones act as relatively rigid rods, and joints often act as a fulcrum or pivotal point.

Three classes of lever are usually recognised:

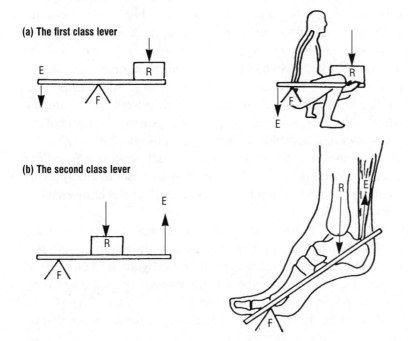

(a) The first class lever

E   R   F

(b) The second class lever

E   R   F

R   E   F

Study tasks
- Highlight the details shown in Figure 3.6.
- Identify the examples given, on yourself, taking care of your own back when lifting.
- Note the effect of increasing/decreasing the distance between 'E' and 'F' (the 'effort/force arm'), and between 'R' and 'F' (the 'resistance/load arm').
- Consider the implications of your findings in the examples shown.

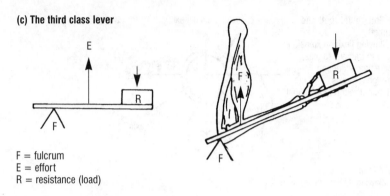

(c) The third class lever

F = fulcrum
E = effort
R = resistance (load)

Figure 3.6 The three main classes of lever

Synovial joint mechanics

Some of the terms used in this book – such as 'spin', 'swing', 'roll' and 'slide' – refer to the terminology of the relatively recent theories of human kinesiology (the study of motion in the human body), and related studies such as arthrokinematics, which deals primarily with the motion of joint surfaces. More familiar terms such as 'flexion/extension', 'abduction/adduction' and 'medial/lateral rotation' will continue to be used whenever possible due to their wide clinical acceptance. However, a brief introduction to some of the concepts of human kinesiology will be given here, and it will on occasion be useful to employ its terminology. Readers who require a more detailed account are advised to consult texts such as Steindler (1955), Barnett *et al.* (1961) and MacConaill and Basmajian (1977). The limitation of relying on terms such as 'flexion' and 'abduction' is that they do not take into consideration the complex additional movements that take place at the articular surfaces of the joints involved, reflecting the varying shapes of these surfaces. For example, flexion at the shoulder is a very different event from flexion of the wrist. In order to be more precise about the movements taking place within a joint it is useful to consider first its mechanical axis.

If the bone moves but the mechanical axis remains stationary, this is known as 'pure' spin. It occurs only if the opposing joint surfaces remain perpendicular to each other (Figure 3.7). Only when the mechanical axis coincides with the long axis or shaft of a moving bone does spin coincide with 'rotation'.

If the mechanical axis moves, it traces a path (as represented by

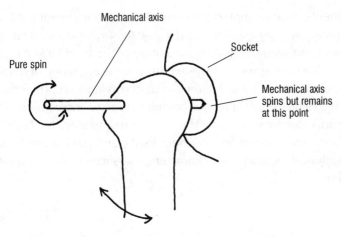

Figure 3.7 The mechanical axis of a long bone showing pure spin

the line of the pen in Figure 3.8), and this movement is referred to as 'swing'.

The shortest path between two points is known as a 'chord', and there is no associated spin when a bone follows this route. A longer path between two points is referred to as an 'arc', and this is accompanied by a degree of spin as well. Therefore, a pure swing (displaying no element of spin) traces the path of a chord (a cardinal swing), whereas an impure swing traces the path of an arc and may also be referred to as an arcuate swing.

The associated spin that occurs automatically in an impure swing may be referred to as 'conjunct' rotation. Most joint move-

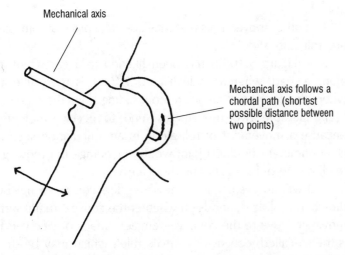

Figure 3.8 The mechanical axis of a long bone showing pure (cardinal) swing

ments, such as shoulder flexion at the glenohumeral joint, are so-called 'impure' movements because they involve varying degrees of spin and swing, reflecting the nature of the articular surfaces.

A section through an ovoid surface (the egg shape of many articular surfaces) shows that the radius of curvature is constantly changing (Figure 3.9). There may be only one area where the mating joint surfaces achieve best fit. This may be an important factor in the stability and lubrication of certain joints, and particularly those with spherical or condylar characteristics, such as the shoulder and knee.

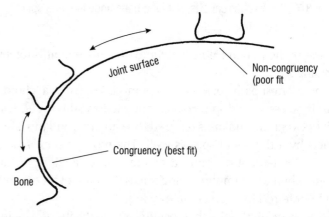

Figure 3.9 Profile through a hypothetical joint showing variations in congruency (fit)

Two other movements that may also take place at joint surfaces are 'roll' and 'slide'.

Roll (Figure 3.10(b)) has been likened to a car wheel rolling along a road, whereas slide (Figure 3.10(c)) is more like a skid, where the car wheels are locked but sliding along the road surface.

In fact the two movements tend to occur simultaneously, because a movement of roll alone in an articulation such as the glenohumeral (shoulder) joint would encourage the moving bone to climb out of its socket and dislocate.

Roll will always occur in the same direction as swing, but the direction of slide depends on whether the convex (male) surface is moving relative to the concave (female) surface or vice versa. This is the so-called 'concave – convex rule', which may be stated as follows:

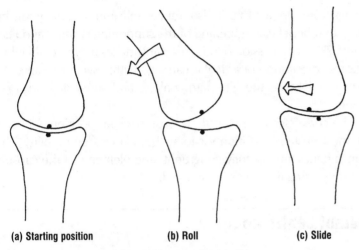

    (a) Starting position   **(b) Roll**   **(c) Slide**

Figure 3.10 Joint surface movements: roll and slide

Roll and slide occur in the same direction if concave moves on convex (Figure 3.11(a)). If convex moves on concave, roll and slide occur in opposite directions (Figure 3.11(b)).

 Study task

- Try to obtain bone specimens of a matching humerus and scapula. Use the glenohumeral spheroidal joint articulations to represent convex and concave surfaces. If specimens are difficult to obtain, simply place the fist of one hand in the cupped palm of your other hand and demonstrate all of the movements described in this section.

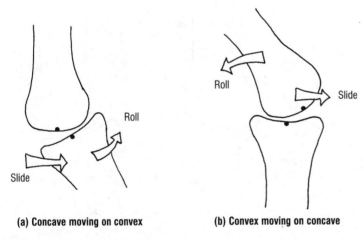

    (a) Concave moving on convex   **(b) Convex moving on concave**

Figure 3.11 The concave–convex rule

Close-packed/loose-packed positions

The approximation of joint surfaces involves a degree of compression. This occurs at some point in the normal range of movement of all joints, and the point of optimum compression of the joint is known as the 'close-packed' position. It implies maximum congruency or best fit of the joint surfaces, and its value as a concept

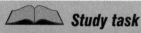 **Study task**

- Work through the following example in order to understand the concepts of instant centre analysis.

springs from the fact that it also implies maximum stability for the joint, with beneficial tightening of the supporting capsule and ligaments. The compression of the articular surfaces may also aid lubrication and nutrition of articular cartilage within synovial joints. An analogy might be the alternate soaking and squeezing dry of a sponge.

The converse is the so-called 'loose-packed' position, which is the most unstable position for joints but allows free mobility. The joint surfaces are not fully congruent, and elements of the capsule and supporting ligaments are relaxed.

Instant centre analysis

This is a technique described by Sammarco *et al.* (1973) that aims to provide a more accurate analysis of movement in joints, particularly in order to differentiate between the components of roll and slide in joint surfaces.

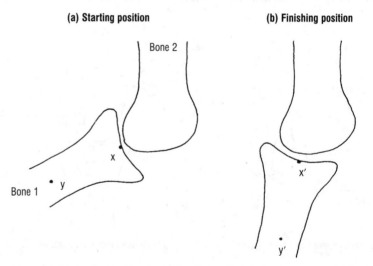

Figure 3.12 Instant centre analysis: stage one

In Figure 3.12(a), bone 1 will move relative to bone 2. Commence by highlighting the two dots on the long axis of bone 1 at point x midway along its articular surface, and at point y at a measured distance along its shaft. Do the same in Figure 3.12 (b), which depicts the position of the bones after movement has taken place.

A radiograph (X-ray/roentgenogram) may be taken before and after movement has taken place, and superimposition of the two radiographs allows the two points to be compared.

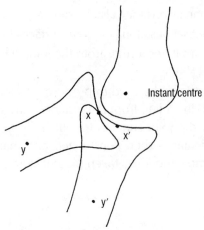

Figure 3.13 Instant centre analysis: stage two (superimposition of the images in Figure 3.12)

Now draw a line to connect points x–x' and y–y' in Figure 3.13, and construct perpendiculars midway along these lines so that they intersect. (In order to help you, the intersection point that forms the 'instant centre' has been identified).

In practice, this exercise may be repeated several times to produce a pathway of instant centres for analysis (Figure 3.14).

(a) Erratic pathway of instant centres in a dysfunctional knee (tibiofemoral joint) before treatment

(b) Smooth pathway of instant centres indicating improvement following treatment

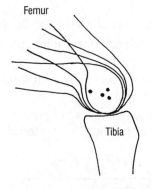
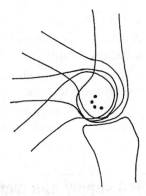

Figure 3.14 Hypothetical examples of instant centre pathways

The value of instant centre analysis is that comparisons may be made of the pathway of a series of instant centres before and after treatment (Figure 3.14), although the two-dimensional nature of the technique means that components such as spin and rotation escape observation.

The following conclusions may also be drawn:

- An instant centre that lies along the joint surfaces indicates roll.
- An instant centre that lies further from the joint line indicates a greater degree of slide.

An instant centre may also be used to produce a 'velocity vector'. This is done by drawing a line from the instant centre to the point of contact between the joint surfaces to which the instant centre relates. A perpendicular is then constructed and marked in the direction in which movement has taken place (Figure 3.15).

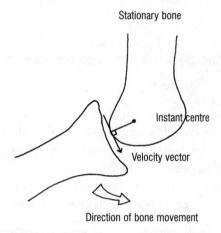

Stationary bone

Instant centre

Velocity vector

Direction of bone movement

Figure 3.15 Determination of a velocity vector

Using a velocity vector the following conclusions may be drawn:

- A velocity vector pointing towards the stationary joint surface indicates compression.
- A velocity vector pointing away from the stationary joint surface indicates distraction.
- A velocity vector pointing along the joint line (as in Figure 3.15) indicates a smooth pathway during the measured interval of motion.

Blood supply and lymphatic drainage

The practical nature of this book means that the vascular (and neurological details) of each joint studied will not be considered in detail. A brief outline of general considerations will be given here; but the reader who requires a detailed exposition of these matters is advised to refer to the appropriate anatomy resources, which are widely available.

Joints receive their blood supply from vascular plexuses that enter the capsule and supply the synovial membrane, eventually anastomosing around the joint. The bones and their epiphyses are supplied separately by branches from the adjacent blood vessels.

The lymphatic drainage is also supplied by plexuses located in the synovial membrane, which then drain to deep lymph nodes.

Nerve supply

The general innervation of joints is from the nerves supplying the muscles that act over them. One of the reasons for this is that a reflex arc may be established, which allows the muscles to protect the joint from injury. For example, a movement such as extension will begin to stretch the opposing surface of the joint capsule. This surface is frequently innervated by the same nerves that supply the flexor muscles (the antagonists). A sudden 'overstretch' will result in a reflex contraction of the antagonistic muscles (in this case the flexors), thus helping to protect the joint from injury.

The nerve supply to joints is strictly afferent (sensory), with the exception of the small vasomotor efferent nerves to the blood vessels. The sensory nerve supply to a joint is designed to provide information about movement, position and forces acting on the body.

Wyke (1967) described four different types of nerve receptor, which he classified as Types I–IV:

Type I

Usually found in the superficial layers of the joint capsule, and thought to monitor joint position and speed of movement. This is important in the hip joint, for example, where sense of position is vital for the maintenance of posture.

Type II

Found deep in joints and in fat pads. Sparser than Type I, they are thought to provide information on rapid changes of movement and pressure.

Type III

Similar to Golgi tendon organs (which are stretch receptors), these play an important role in ligaments. They are thought to monitor fast movements likely to cause the joint to exceed normal range. As such they may help to initiate a reflex contraction of antagonistic muscles in order to protect the joint from overstretch.

Type IV

Found widely in joint tissue, and probably sensitive to chemical stimulation (for example from inflammatory responses).

Chapter 4

The structure of the vertebral column: an introduction

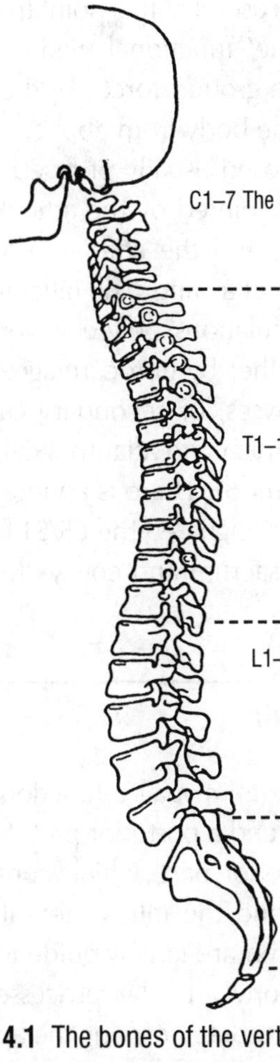
C1–7 The cervical spine
T1–12 The thoracic spine
L1–5 The lumbar spine
The sacrum
The coccyx

Figure 4.1 The bones of the vertebral column

Study tasks

- Shade the different regions of the vertebral column in separate colours.
- Highlight the details shown.
- Think of as many functions of the vertebral column as you can.

The vertebral column is a segmented flexible pillar (once irreverently described as 'a wobbly pile of bones') which consists of seven cervical (neck) vertebrae designated C1–7, 12 thoracic (chest and rib) vertebrae designated T1–12 and five lumbar (low back) verte-brae designated L1–5. There is also a sacrum (literally 'sacred' bone), which consists of five fused vertebral segments (S1–5), and a coccyx (tail bone) consisting of four tiny fused segments.

In functional terms the vertebral column should also include the occipital condyles which transfer the weight of the head to the uppermost cervical vertebra (C1). This is a highly specialised verte-bra aptly named the 'atlas' after the titan in classical mythology whose role was to support the world on his shoulders.

The second cervical vertebra (C2) is also a specialised structure termed the 'axis', since its function is to rotate the head.

The articulations between the sacrum and the pelvis will be con-sidered separately (Chapter 9), but it is useful at this point to appre-ciate the significance of the 'keystone' functional wedge of the sacrum, since this helps to absorb the ground forces of the lower limbs from below and the weight of the body from above.

The sacrum and coccyx are composed usually of fused bones, but the rest of the vertebral column is jointed by the articular sur-faces between the occipital condyles and the atlas (the atlanto-occipital joints) and between the the atlas and axis (atlanto-axial joints) and by all the individual articulations between vertebrae C2–S1. These are all held firmly together by fibrocartilage, which forms the intervertebral discs (symphyses, or secondary cartilagi-nous joints), except at the atlanto-occipital and atlanto-axial joints, where no discs are present (see Chapter 5). There is an interverte-bral disc between L5 and the sacrum (designated the L5/S1 or lum-bosacral disc); and one between the sacrum and coccyx (forming the sacrococcygeal joint).

The structure of the vertebrae

A sagittal view of a vertebra (Figure 4.2(a)) reveals a functional unit which can be divided into an anterior and a posterior part. Most of the body weight is borne by the anterior part, which consists of appropriately large vertebral bodies and the intervertebral discs. The movements of the vertebral column are largely guided by the posterior part, which includes the bony articular processes that interlock with those of the adjacent vertebrae, forming the apophy-seal (zygapophyseal) joints. Once two vertebrae are structurally linked by an intervertebral disc and supporting ligaments (see below), the resultant functional unit may be referred to as a 'motion segment' (Figure 4.2(a)).

(a) Vertebral motion segment

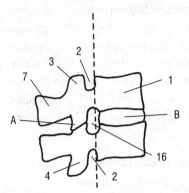

Posterior part
(guides movement
and protects spinal
cord)

Anterior part
(supports weight)

(b) Typical cervical vertebra

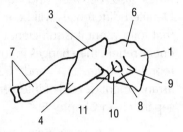

(c) Typical thoracic vertebra

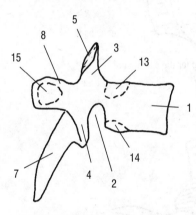

(d) Typical lumbar vertebra

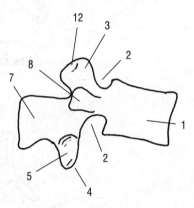

 1 Body of vertebra
 2 Vertebral notch (superior and inferior together create the intervertebral foramen)
 3 Superior articular process
 4 Inferior articular process
 5 Facet for apophyseal joint (A)
 6 Uncinate/unciform process (cervical vertebrae only)
 7 Spinous process (bifid in cervical spine, except C7)
 8 Transverse process (not shown in (A))
 9 Anterior tubercle of cervical transverse process
10 Posterior tubercle of cervical transverse process
11 Foramen transversarium (for vertebral artery – cervical vertebrae only)
12 Mamillary process (lumbar vertebrae only)
13 Demi-facet for rib head (thoracic vertebrae only)
14 Demi-facet for rib head (thoracic vertebrae only)
15 Facet for tubercle of rib (thoracic vertebrae only)
16 Intervertebral foramen (for nerve root)
 A Apophyseal (zygapophyseal) joint (synovial, plane)
 B Intervertebral disc (symphysis)

Figure 4.2 Sagittal views of vertebral bone features

In spite of similarities, there are also noticeable structural differences between cervical, thoracic and lumbar vertebrae, and sometimes also between individual vertebrae within these groups. Detailed differences will be examined in the appropriate chapters that follow, but it is sufficient to state here that the structural differences reflect the changes in function which occur through the vertebral column. C1 and C2 present a different specialised arrangement devoid of intervertebral discs, which will be examined separately in Chapter 5.

Study tasks

- Highlight the features of the vertebra shown in Figure 4.3(a).
- Match the following functions of a vertebra with the features on the diagrams: (i) support; (ii) shock absorption; (iii) protection; (iv) leverage; (v) metabolic functions (including manufacture of blood cells from bone marrow).

(a) Schematic oblique view of a (lumbar) vertebra

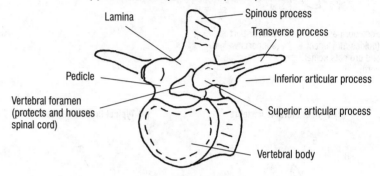

(b) Protective housing for the spinal cord

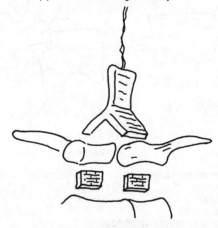

(c) Trabeculae (lines of force) pass through internal spongy bone, adding strength to the structure (sagittal section)

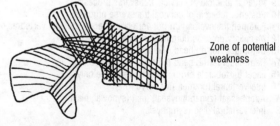

Figure 4.3 Main features of a typical (lumbar) vertebra

In oblique or horizontal views, the anterior and posterior parts look rather different. The posterior part functions as a bony ring or neural arch, which surrounds and protects the spinal cord. This protective arrangement may be compared imaginitively with the walls and roof of a house (Figure 4.3(b)), with the body of the vertebra forming the foundations.

The segmental nerve roots from the spinal cord emerge through notches between the pedicles (viewed sagittally), forming the intervertebral foramina between adjacent vertebrae (Figure 4.2(a)). The bodies of the vertebrae are further strengthened by 'force lines' known as trabeculae (singular: trabecula) which pass through the inner spongy (cancellous) bone where red blood cells are manufactured. These are depicted in Figure 4.3(c), which also shows that there is a potential point of weakness in the anterior part of the vertebral body. This may collapse in certain bone pathologies, producing a flexion deformity in the vertebral column. The outer casing of a vertebra, in common with other bones, consists of compact or 'cortical' bone which is more resistant.

The joints and ligaments

Examination of a flexible plastic spine, or matching bone specimens, reveals how the bony structure of adjacent vertebrae allows the posterior articular processes of the motion segments to interlock. These form plane synovial joints called apophyseal (zygapophyseal) joints, which will be discussed in the chapters that follow. The anterior vertebral bodies are linked by fibrocartilaginous intervertebral discs (symphyses). Details of the structure of these are given in Chapter 7.

The entire vertebral column is linked by ligaments and supporting muscle. The ligaments contain proteins such as collagen (for strength) and elastin (for elasticity) in varying proportions, and also contain stretch receptors to monitor the various forces that are exerted on the vertebral column. The anterior longitudinal ligament resists excessive extension, whereas most of the other ligaments are positioned posteriorly to prevent excessive flexion.

Some ligaments, such as the ligamentum flavum, may also contain a certain amount of contractile tissue like a muscle, which probably makes their protective functions more effective. The vertebral ligaments may be conveniently classified into those attaching to the anterior part of the motion segment and those attaching to the posterior part.

The cervical and thoracic parts of the spine contain specialised ligaments, which will be dealt with in later chapters.

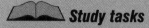 **Study tasks**

• Shade the ligaments in separate colours, distinguishing between the anterior and posterior ligaments.
• Highlight the names of the ligaments and the features shown.

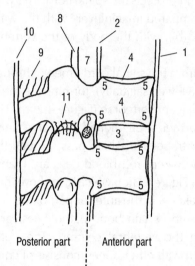

1 Anterior longitudinal ligament
2 Posterior longitudinal ligament
3 Intervertebral disc
4 Hyaline cartilage vertebral end plate
5 Ring epiphyses (growth zones in immature bone)
6 Nerve root in intervertebral foramen
7 Vertebral canal (spinal cord)
8 Ligamentum flavum
9 Interspinous ligament
10 Supraspinous ligament
11 Apophyseal joint (with enclosing fibrous capsule)

Posterior part   Anterior part

Figure 4.4 Ligaments and main features of a (lumbar) vertebral motion segment (saggital view)

Movements

These are so intimately connected with the speciality of function in the cervical, thoracic and lumbar spine that they will be examined individually in each chapter as appropriate. Nevertheless, it may be useful to review the terms that will mainly be used to describe movements of the vertebral column in this book. Conventional clinical terminology refers to the terms 'flexion', 'extension', 'lateral flexion/sidebending', 'rotation' ('axial rotation') and 'circumduction'.

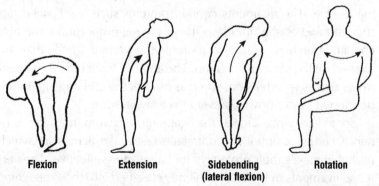

Flexion   Extension   Sidebending   Rotation
                        (lateral flexion)

Figure 4.5 Movement of the vertebral column (conventional clinical terminology)

These terms are usually adequate to describe gross movements, but have limitations when describing movements at joint surfaces, as already discussed in Chapter 3. For this reason, a more rigorous terminology which refers to coordinate axes (X, Y and Z, in Figure 4.6) is employed in the literature of kinesiology, and is included here in simplified form for reference purposes. This classification includes sliding movements referred to as 'translation', as well as compression and distraction. Under this system it is possible to refer to six degrees of freedom of movement.

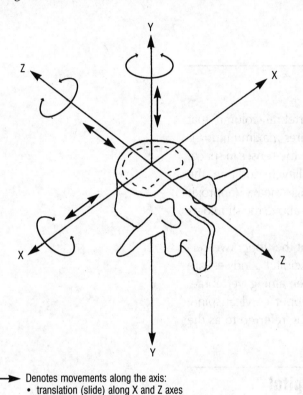

Denotes movements along the axis:
- translation (slide) along X and Z axes
- compression/distraction along Y axis

Denotes rotatory movements
- flexion/extension around X axis
- sidebending (lateral flexion) around Z axis
- rotation around Y axis

Figure 4.6 The three coordinate axes of movement in a vertebral motion segment (showing six degrees of freedom)

Chapter 5

The cervical spine

Introduction

The cervical spine is the section of the vertebral column that directly supports the head, and therefore requires maximum possible freedom of movement in order to orientate the senses in space. The cost of this freedom is a reduction in stability. In addition, the weight of an average adult head is surprisingly heavy (approximately 5–7 kg) and the cervical spine also rises above the shoulders in a somewhat vulnerable way.

Function dictates structure to the extent that the upper two cervical vertebrae (functionally linked with the occipital condyles) are completely different in structure from the remaining vertebrae, reflecting the specialised movement of the upper cervical spine. The remaining cervical vertebrae (C3–7) will be referred to as the 'lower cervical spine'.

The upper cervical spine (sub-occipital region/craniovertebral junction)

The bones

The occipital condyles and the atlas (C1)

The uppermost vertebra, the atlas, supports the head by accepting the occipital condyles which rest on its superior articular facets. The vertebra is named after the titan in classical mythology who carried the world on his shoulders. The most striking feature of this vertebral ring of bone which supports the head is the obvious lack of a vertebral body. The apparently missing body has in fact become detached and fused to the vertebra below (the axis), allowing the head and atlas to rotate as one unit around the tooth-like pivot so formed, which is variously termed the 'dens' (Latin 'tooth') or 'odontoid process' (Figure 5.2). This arrangement also explains the absence of an intervertebral disc between C1 and C2.

(a) Occipital condyles and atlas (C1) (posterior view – opened out

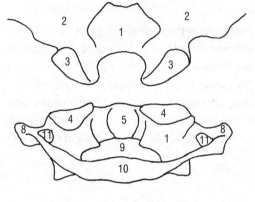

(b) Atlas (C1) (superior view)

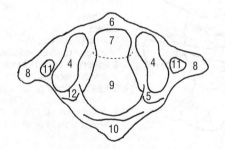

1 Foramen magnum (for spinal cord)
2 Occipital bone
3 Occipital condyles
4 Superior articular facet of atlas
5 Facet for dens on anterior arch of atlas
6 Anterior arch and tubercle of atlas
7 Space for dens
8 Transverse process of atlas
9 Vertebral foramen for spinal cord
10 Posterior arch and tubercle (spinous process) of atlas
11 Foramen transversarium

- - - - - Outline of transverse ligament of atlas

Study tasks

- Obtain a bone specimen of the atlas and note the features shown in Figure 5.1.
- Palpate the transverse processes of the atlas on yourself by gently locating the gap between your mastoid process (behind the ear) and the angle of the mandible (jaw). The transverse processes feel like resistant masses.
- Also notice in your palpation that you cannot feel the spinous process of the atlas. The first palpable spinous process is that of the axis (C2).

Figure 5.1 The occipital condyles and atlas (C1)

The axis (C2)

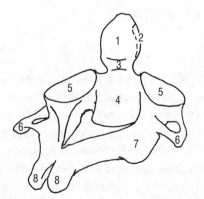

1 Dens (odontoid process)
2 Facet for articulation with anterior arch of atlas
3 Groove for transverse ligament of atlas
4 Body
5 Superior articular facet
6 Posterior tubercle of transverse process
7 Inferior articular process
8 Bifid spinous process

Figure 5.2 The axis (C2) (oblique posterior view)

The axis is easily recognised by the tooth-like dens, which has a small facet on its anterior surface around which the anterior arch of the atlas rotates. Notice also that the spinous process is larger and looks more powerful than that of the atlas. It is bifid (split), like vertebrae C3–6, which allows attachment for the ligamentum nuchae (see Figure 5.9) as well as the rectus capitis posterior major muscle of the sub-occipital group. This makes the spinous process prominent and easy to palpate.

The transverse processes are represented solely by a posterior tubercle for the middle scalene muscle; the anterior scalene muscle is attached to C3. The superior articular surfaces (facets) lie in an approximately horizontal plane for articulation with the inferior articular facets of the atlas. The inferior articular facets of the axis are set back and at an angle of approximately 45° in a semi-coronal plane which is at the same angle as the cervical vertebrae below.

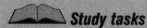

Study tasks

- Obtain a bone specimen of the axis and identify the features shown in Figures 5.2 and 5.3.
- Shade the occipital condyles, atlas and axis in separate colours.
- Palpate the spinous process of the axis on your own neck, noting the hollow formed immediately above by the minimal spinous process of the atlas.

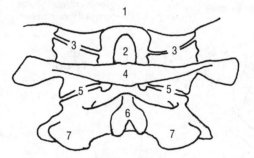

1 Occiput
2 Dens
3 Articular surfaces of occipital condyles and atlas
4 Posterior arch and tubercle (spinous process) of atlas
5 Lateral articular surfaces of atlas and axis
6 Spinous process of axis
7 Inferior articular process of axis

Figure 5.3 The occipital condyles, atlas (C1) and axis (C2) (posterior view)

The joints and ligaments

CO/C1: the atlanto-occipital joints

The convex surfaces of the occipital condyles are slightly larger than the concave receptor surfaces on the superior surface of the atlas. The latter are somewhat kidney-shaped, and angled obliquely. These joints are lined by a synovial membrane which may communicate with a bursa between the dens and transverse ligament of the atlas (Figure 5.4(b)). The atlas, axis and occipital condyles are so intimately connected that all relevant ligaments will be described here together.

(a) Deep ligaments (posterior view)

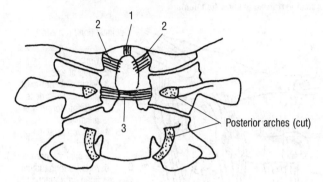

Posterior arches (cut)

(b) The cruciform ligament transforming transverse ligament into a 'cross' (posterior view)

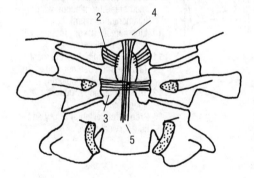

(Key (a–c))
1 Apical ligament between dens and anterior border of foramen magnum
2 Alar ligament between dens and medial occipital condyles
3 Transverse ligament of atlas retaining dens in contact with anterior arch of atlas
4 Superior longitudinal band of cruciform ligament linking transverse ligament and occiput
5 Inferior longitudinal band of cruciform ligament linking transverse ligament and body of axis
6 Tectorial membrane attached to occiput and becoming posterior longitudinal ligament

(c) The tectorial membrane covering the cruciform ligament (posterior view)

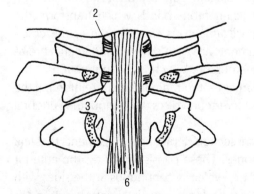

Figure 5.4 The atlanto-occipital and atlanto-axial joints

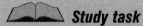

(d) Median sagittal section through occipital bone and upper cervical vertebrae to show ligaments

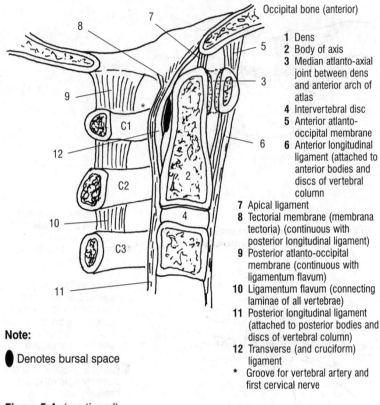

Occipital bone (anterior)

1 Dens
2 Body of axis
3 Median atlanto-axial joint between dens and anterior arch of atlas
4 Intervertebral disc
5 Anterior atlanto-occipital membrane
6 Anterior longitudinal ligament (attached to anterior bodies and discs of vertebral column
7 Apical ligament
8 Tectorial membrane (membrana tectoria) (continuous with posterior longitudinal ligament)
9 Posterior atlanto-occipital membrane (continuous with ligamentum flavum)
10 Ligamentum flavum (connecting laminae of all vertebrae)
11 Posterior longitudinal ligament (attached to posterior bodies and discs of vertebral column)
12 Transverse (and cruciform) ligament
* Groove for vertebral artery and first cervical nerve

Note:

● Denotes bursal space

Figure 5.4 (continued)

Figure 5.4(a) shows the ligaments that support and restrain the dens. The transverse and alar ligaments are especially important, and frequent referral will be made to them. The apical ligament holds the tip of the dens in place. Figure 5.4(b) shows the addition of superior and inferior ligamentous bands which transform the transverse ligament into a ligamentous cross known as the cruciform ligament. These ligaments are then covered and in effect reinforced by the tectorial membrane, which is a continuation of the posterior longitudinal ligament – the main ligament reinforcing the posterior surface of the intervertebral discs throughout the vertebral column.

Figure 5.4(d) is a sagittal section depicting the ligaments that link the atlas and occipital bones. These consist of, firstly, the anterior atlanto-occipital membrane, which is continuous at the sides with the apophyseal capsules of the atlanto-occipital joints. It is strengthened centrally by the anterior longitudinal ligament, which reinforces the anterior surface of all intervertebral discs. The posterior atlanto-occipital membrane connects the posterior margin of the foramen magnum to the posterior arch of the atlas. It passes over the groove on the atlas for the vertebral artery and first cervical nerve, and may ossify at this point. **For this reason, extreme care should be taken when these joints are manipulated.**

C1/C2: the atlanto-axial joints

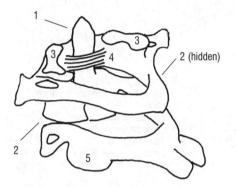

1 Median atlanto-axial joint (synovial, pivot)
2 Lateral atlanto-axial joints (synovial, plane)
3 Superior articular facet of atlas
4 Transverse ligament of atlas
5 Inferior articular process of axis

 Study tasks

- Colour the atlas and axis in separate colours in Figure 5.5.
- Refer back to Figure 5.4 and note the ligaments that connect the atlas and axis and those that connect the axis and occiput.

Figure 5.5 The atlanto-axial joints (posterior oblique view)

There are three articulating surfaces at the atlanto-axial joint, consisting of the median pivot joint and the two lateral apophyseal joints. The dens is held in place by the transverse ligament. This is lined by articular cartilage, and a bursa is present. The hyaline cartilage that lines the lateral joints uniquely converts them into two opposing convex surfaces which are angled slightly obliquely.

Movements and muscles

There are are wide variations in reported ranges of movement in the cervical spine due to difficulties in accurate measurement (Worth, 1994). Age differences complicate matters even further. Therefore figures of ranges of movement quoted here are probably best regarded as a general guide only.

All movements described below relate to those traditionally used in clinical examination.

C0/C1: the atlanto-occipital joints

Flexion/extension. The predominant functional movement at this joint is that of flexion/extension, popularly referred to as 'nodding', with ranges of movement between 15 and 50° being reported (Worth, 1994).

Study tasks

- Palpate the relatively small range of flexion/extension on yourself by resting your thumbs on the transverse processes of your atlas (see study tasks on p. 31) with the palmar tips of your fingers on the sub-occipital muscles on each side at the top of your neck. Perform small nodding movements without moving your neck below your palpating fingers.
- With reference to Figure 5.4(d), consider which ligaments and structures restrict excessive flexion and extension at the atlanto-occipital joints.
- Muscle check.

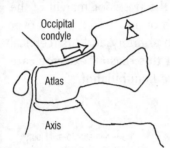

(a) Flexion

Occipital condyle

Atlas

Axis

Movement produced by:
- longus capitis
- rectus capitis anterior

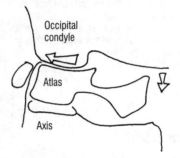

(b) Extension

Occipital condyle

Atlas

Axis

Movement produced by:
- rectus capitis posterior major and minor
- obliquus capitis superior
- semispinalis capitis
- splenius capitis
- trapezius (upper fibres)

Figure 5.6 Flexion/extension of the atlanto-occipital joints (median sagittal section)

Note the translation (slide) that occurs in the opposite direction to the main movement. During extension from full flexion there is also an equivalent degree of anterior translation of the atlas relative to the axis, which is limited by the transverse ligament of the atlas.

Sidebending (lateral flexion). Sidebending specifically at this joint is probably only of the order of 5–8° on each side. Some authorities, for example Steindler (1955) and Kapandji (1974), regard this movement as a lateral tilt accompanied by conjunct rotation to the opposite side. This may be necessary to allow the ipsilateral occipital condyle to become congruent and stable on the superior facet surface of the atlas below.

Excessive sidebending at this joint is highly undesirable because it exposes the medulla oblongata (upper spinal cord) to the risk of injury, and for this reason the movement is strictly limited by the alar ligaments (Figure 5.4(a)). These run from each side of the upper dens to the medial sides of the occipital condyles. Acting together, they check flexion and rotation at the atlanto-occipital joint, and individually limit sidebending; they relax in extension of the head. Jirout (1973) suggests that sidebending at the atlanto-occipital joint is accompanied by rotation at the atlanto-axial joint. This is caused by the insertion of the alar ligament on the dens, which produces rotation of the axis towards the side of lateral flexion. The muscles that produce sidebending at the atlanto-occipital joints are the rectus capitis lateralis, semispinalis capitis, splenius capitis, sternocleidomastoid and the upper fibres of the trapezius (Williams, 1995); the sub-occipital muscles assist.

Rotation.   The configuration of the articular surfaces appears to limit rotation at the atlanto-occipital joint, and many authorities discount it altogether (White and Panjabi, 1978b). Observations by Penning and Wilmink (1987) using computerised axial tomography seem to confirm the view of Kapandji (1974) and others that a small degree of sidebending to the opposite side always accompanies rotation at the atlanto-occipital joints.

C1/C2: the atlanto-axial joints

This is the most mobile segment in the vertebral column, the most significant movement being rotation.

Flexion/extension.   These are minimal movements at this segment and are strictly limited to a slight permissive slide of the anterior arch of the atlas on the dens, restrained by the transverse ligament (Figure 5.5).

Sidebending (lateral flexion).   This is also strictly limited by the proximity of the dens to the lateral mass of the atlas, and tends to merge into rotation to the same side.

Rotation.   It is usually accepted that the atlanto-axial joints contribute approximately 50% of total cervical rotation, which is about 90° to each side. Thus, about 45° of axial rotation is available on each side at the atlanto-axial joints. As the anterior arch pivots on the dens, one inferior lateral articular facet of the atlas slides back

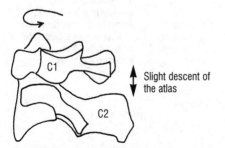

C1
Slight descent of the atlas
C2

Movement produced by:
• obliquus capitis inferior
• rectus capitis posterior major
• splenius capitis (all on the same side as rotation)
• sternocleidomastoid (on the opposite side to rotation)

Figure 5.7 The atlanto-axial joints: rotation to the left (oblique lateral view)

> ### Study task
>
> • Palpate this movement of sidebending (and the integrity of the alar ligaments) by placing the tips of your fingers on the spinous process of the axis and sidebending your head at the atlanto-occipital joint. The spinous process of the axis will be felt to move slightly to the side opposite sidebending, which is indicative of the vertebral body of the axis moving towards the side of lateral flexion.

Study tasks

Important preliminary safety note:
The integrity of the upper cervical spine may be damaged by impact injuries or weakened by systemic rheumatoid disorders. The proximity of the vertebral artery must also be borne in mind. For this reason a proper case history should be taken before an examination of the upper cervical spine is undertaken, and even then the examination should be made only under qualified supervision. Do not proceed with the tasks outlined if the subject experiences adverse symptoms.

• Place your index finger tips on the transverse processes of your atlas and gently rotate your head from side to side. You should be able to feel the transverse processes of the atlas as they move over the lateral surfaces of the axis.
• Palpate also the slight sidebending to the contralateral side at the atlanto-occipital joint which takes place during this movement.
• Repeat these movements on a supine colleague and note the increase in range obtained if extension and sidebending to the opposite side are added.
• Muscle check.

on the surface of the reciprocating superior facet of the axis on one side, while the other superior facet slides forward (Figure 5.7). The shape of the articular hyaline cartilage on the lateral facet surfaces of both atlas and axis renders both surfaces convex, so that the atlas drops slightly on the axis in a screw-like motion of vertical translation which Kapandji (1974) suggests may be as much as 2–3 mm. This movement is eventually limited by the alar ligaments on each side, but the slight descent of the atlas just described has the effect of delaying the tension in the apophyseal joint capsules and alar ligaments; thus a little more rotation is possible. This may explain why it is possible to increase the range of rotation at this joint by adding extension with sidebending on the contralateral side, which slackens both the capsule and alar ligaments. (**Safety note: do not proceed with or repeat these movements if dizziness is experienced.**)

The lower cervical spine (C3–7)

The lower cervical spine is regarded as the region from the inferior surface of the C2 vertebra to the inferior surface of C7 (C7–T1 being the cervicothoracic junction; see Chapter 8). It is convenient to describe this area as functionally separate, but it should not be forgotten that both the upper cervical spine and the upper thoracic spine must be considered in a proper functional clinical assessment.

The bones

(a) Superior view

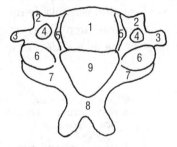

1 Vertebral body
2 Anterior tubercle of transverse process
3 Posterior tubercle of transverse process
4 Foramen transversarium
5 Uncinate process
6 Superior articular facet
7 Inferior articular facet (b); process only shown in (a)
8 Bifid spinous process
9 Vertebral foramen

(b) Lateral view

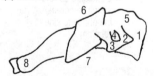

Figure 5.8 A typical lower cervical vertebra (C3–5)

C3–7 are collectively referred to as the 'lower cervical spine', but only C3–5 should be regarded as typical. The anterior tubercle of C6 is larger, known as the carotid tubercle, and this lies posterior to the common carotid artery, hence its name. C7 has a non-bifid spinous process with tubercle for the attachment of the ligamentum nuchae (Figure 5.9) and the slightly sloping process renders it transitional to the thoracic vertebrae below. It also possesses a foramen transversarium, but the vertebral artery in the neck actually enters at C6 level. The presence of this foramen for the transmission of the vertebral artery and venous and sympathetic nerve plexuses is a distinguishing feature of the cervical vertebrae.

The vertebral bodies are relatively small (supporting only the weight of the head) and broad but concave transversely on their superior surface. This gives rise to the raised lateral lips known as 'uncinate' (unciform) processes, which in turn give rise to occasional synovial joints, sometimes referred to as 'uncovertebral' joints or the 'joints of Luschka'.

The inferior surface of the bodies has an anterior lip that overhangs the vertebra below, which possesses a bevelled superior surface to accommodate this, and the intervertebral disc.

The articular facets are angled in a semicoronal plane of approximately 45° so that the articular processes stack to form a lateral articular pillar which is palpable. The apophyseal facet joints between these surfaces therefore appear to guide rather than limit movement (compared with the lumbar spine, where a sagittal orientation of the facets serves to limit rotation; see Chapter 7). The uncinate processes may also guide movements, but probably also have a stabilising role.

Notice also the relative size of the vertebral foramen. Its generous dimensions allow adequate room for the neurovascular structures of the spinal cord during versatile movements of the head and neck.

It is reasonable to assume that the structure of the cervical vertebral column is designed to provide stability with mobility as well as support and protection.

The joints and ligaments

The intervertebral discs

These are present between vertebrae C2–S1, and in the cervical and lumbar areas tend to be slightly thicker anteriorly, which helps

---

Clinical Note

The presence of the foramen transversarium, together with the grooved surface for the spinal nerve roots, weakens the transverse process and causes a predisposition to fractures in the cervical spine.

Study tasks

- Highlight the bone features shown in Figure 5.8.
- Match the features shown in Figure 5.8 with the following terms:
  (i) weightbearing; (ii) support; (iii) protection; (iv) mobility; (v) strength.
- Identify all named features on a bone specimen.

Study tasks

- Highlight the names of all features shown in Figure 5.9 and shade the ligaments in colour.
- With reference to a plastic flexible spine if necessary, try to identify by palpation the following structures on a supine colleague:

  - spinous processes (linked by the ligamentum nuchae);
  - the lateral articular pillar of apophyseal joints.

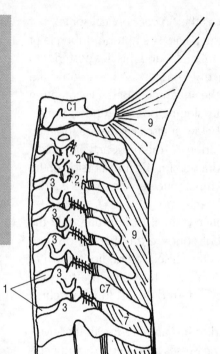

1 Intervertebral disc (symphysis)
2 Apophyseal joint and capsule (synovial, plane)
3 Uncinate process with uncovertebral joint (of Luschka) (synovial, plane)
4 Anterior longitudinal ligament
5 Posterior longitudinal ligament
6 Ligamentum flavum
7 Interspinous ligament
8 Supraspinous ligament
9 Ligamentum nuchae

Figure 5.9 The joints and ligaments of the lower cervical spine in the context of the cervical region as a whole (lateral view)

Clinical Notes

It is of interest that, however ill-defined, these structures have been long implicated clinically (especially by manual therapists) in the acute 'locked neck' syndrome, where the patient suffers muscle spasm for no apparent reason, and this may help to explain why manipulation often seems to be an effective treatment for this condition (Mercer, 1994).

to reinforce the lordosis found in these regions. The ratio of disc to vertebral body height is 2:5 (Kapandji, 1974); the height of the discs is greater in the early years of life, diminishing with age due to a reduction in proteoglycan content (see Chapter 7). With advancing age, this tends to result naturally in diminishing cervical mobility.

The apophyseal joints and inclusions

The presence of inclusions* in synovial joints generally is common. They fall into three categories: intra-articular fat pads (for example, also present in the elbow, hip and knee); fibro-cartilaginous menisci (for example, also present in the knee and shoulder joints); and capsular rims (also found in the hip and shoulder joints). Some of these structures have been identified in cervical apophyseal joints (Mercer, 1994) although research evidence is sparse at present. Small fibro-adipose inclusions in apophyseal joints have been termed 'meniscoids' and their presence may provide greater congruency or help in the transmission of forces.

The uncovertebral joints

The presence of the uncinate processes is a consistent feature in cervical vertebrae and appears to reinforce lateral stability. The presence of these small synovial (plane) joints between the articulating surfaces is less consistent, and embryological studies have suggested that uncovertebral joints (joints of Luschka) are not present at birth and may develop later in response to the process of ageing and degenerative change (Hirsch *et al.*, 1967). This would presumably explain the inconsistency.

Movements and muscles

All degrees of freedom of movement are possible in the lower cervical spine, and together with the upper cervical segments the cervical spine is the most mobile region in the vertebral column, allowing the head maximum range of motion – which is greatest in childhood. Values indicating expected range of movement will be quoted but are highly variable in available sources, reflecting the difficulties of accurate measurement, as well as the variability between individuals. As stated previously, such values should be taken as a guide only.

Cervical curves

Observation of normal erect cervical posture, and inspection of radiographs of the same, reveals that there are two curves present in the cervical spine: a longer lordotic curve from C3 to C7 (Greek *lordos* = 'bent backwards'), and a reversal of this in the upper cervical spine, which is necessary in order to keep the eyes level with the horizon. This reflects the fact that the upper and lower cervical spine may operate independently of each other.

Flexion

(angle may be measured from the horizontal 'plane of bite', 0–75°)

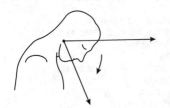

Movement produced by:
* longus colli
* scalenus anterior, medius, and posterior
* sternocleidomastoid

Figure 5.10 Flexion of the lower cervical spine (detail of motion segment)

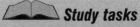 **Study tasks**

* Observe several colleagues and notice the lordotic curve in the lower cervical spine.
* Palpate the slight flexion or 'straightening' of the upper cervical spine.

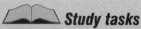

Study tasks

- With reference to Figures 5.9 and 5.10 consider which structures and ligaments prevent excessive flexion.
- Muscle check.
- Carefully extend your own upper cervical spine (palpate this), but then gradually flex the lower cervical spine. Reverse this with upper cervical flexion and lower cervical extension, and you will have demonstrated that these two areas can flex/extend independently of each other.
- Observe how the degree of upper and lower cervical flexion varies according to the order in which the movements are performed. Try to flex the upper cervical spine first (tuck your chin in) and then proceed to curl your neck forwards into full flexion. Palpate this movement as it occurs in the lower cervical spine. Now repeat flexion, but do not consciously flex the upper cervical spine first. More flexion occurs in the lower cervical spine.
- As a result of these findings consider the various anatomical structures which will be stretched by these movements, and try to explain the differences.
- Examine a flexible plastic spine to see what happens to the intervertebral foramina during flexion, noting the clinical significance of your findings for the emerging nerve roots. It is important clinically to examine upper and lower cervical flexion separately, as well as the gross movement.

The upper vertebra tilts and slides forward aided by the bevelled upper surface of the vertebral body below and the oblique plane of the facet joints.

The anterior part of the intervertebral disc is compressed, and the nucleus pulposus is forced posteriorly. The posterior part of the disc and the posterior ligaments are stretched.

Extension

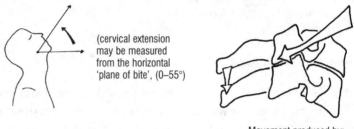

(cervical extension may be measured from the horizontal 'plane of bite', (0–55°)

Movement produced by:
- erector spinae
- splenius cervicis
- semispinalis cervicis

Figure 5.11 Extension of the lower cervical spine (detail of motion segment)

The upper vertebra tilts and slides down guided by the oblique facet planes of the vertebra below. The posterior part of the intervertebral disc is compressed and the nucleus pulposus is forced to move anteriorly. The anterior part of the disc and associated ligament is stretched.

Sidebending (lateral flexion) and rotation

Rotation mainly produced by:
- muscles that rotate the head (Figure 5.7)
- splenius cervicis (to the same side as the movement)
- scalenus anterior
- longus colli (inferior fibres)
- semispinalis cervicis
- rotatores
- multifidus
- trapezius (upper fibres)
  (all acting on the opposite side to the movement)

Sidebending produced mainly by:
- muscles that sidebend the head on the neck (p. 37)
- sternocleidomastoid
- longus colli
- scalenus anterior, medius and posterior
- levator scapulae
- intertransversarii
- erector spinae
  (all acting on the same side as the movement)

Figure 5.12 Sidebending and rotation of the cervical spine (generalised anterior oblique view)

In the lower cervical spine, sidebending and rotation always occur together, and to the same side. This is due to the shape and orientation of the articular surfaces of the apophyseal joints. On the ipsilateral side, the lower articular facet slides down and back, while on the other side the lower facet slides upwards and forwards. This produces sidebending with rotation to the same side, and is a constant feature whether the neck is in flexion or extension. The angle of the articular surfaces (approximately 45°) results in a load-bearing bony articular column which ensures that the apophyseal joints are always in control of lower cervical movements.

Clinical Note

Whiplash injuries occur when the neck is flung violently into extension and then flexion, usually as the result of a rear-end vehicle collision. It will have been noted that there is no bony resistance to flexion in the cervical spine. Injury is often caused to the restraining ligaments or in extreme cases results in anterior dislocation of the superior vertebrae. Posterior structures may be injured by the initial extension phase of impact.

Clinical Note

Rotation and sidebending increases the size of the intervertebral foramina on the contralateral side. This can be used to relieve pressure on an irritated nerve root, especially if combined with flexion and manual traction. This anatomical finding underlies a number of cervical manipulation techniques.

Upper and lower cervical movements

Upper and lower cervical movements work together in a complex but logical manner. For instance, it has already been demonstrated that flexion and extension can occur independently in the upper

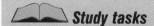

Study tasks

- With reference to Figures 5.9 and 5.11 consider which structures and ligaments prevent excessive extension in the cervical spine.
- Muscle check.
- Examine a flexible plastic spine to see what happens to the intervertebral foramina during extension, noting the clinical significance of this for the emerging nerve roots.

Study tasks

- Flex your neck and palpate the spinous processes of the lower cervical spine. Sidebend your neck and notice the rotation of the spinous processes to the opposite side. This means that the vertebral bodies have rotated to the same side as the sidebending.
- Observe and palpate these movements on a seated colleague.
- With reference to a flexible plastic spine examine the movements of sidebending and rotation, noting the movements of the articular surfaces. Observe the intervertebral foramina in the lower cervical spine during rotation and sidebending movements.
- Muscle check.

and lower cervical spine. Sidebending/rotation movements must also operate independently for the simple reason that it is not always appropriate that the head should be forced into a position of rotation when the neck is sidebent.

Independence of movement may also be used by the body to compensate for postural malalignment. Figure 5.12 shows what happens in full rotation of the head and neck to the right. Logically we can assume that the movement is desirable only because it is necessary to rotate the head (i.e. the senses).

C3–7 are in rotation with sidebending to the right. These vertebrae also exhibit a variable degree of extension due to the cervical lordosis. The atlanto-axial joint is rotated to the right but the atlanto-occipital joint will be sidebent to the left and slightly flexed. This keeps the eyes level with the horizon.

How, one might ask, is it then possible to sidebend the head and neck without automatic rotation of the head taking place? This is possible because the upper and lower cervical spine are able to operate independently. Thus, in full sidebending to the right, C3–7 will be in sidebending/rotation to the right as above, but this time the atlanto-axial joint will be rotated to the left, with the atlanto-occipital joint sidebent to the right and with a degree of flexion.

Clinical Note

The independence of movement between the upper and lower cervical spine in rotation/sidebending also allows the body to compensate for postural malalignment. Any structural or functional rotation or sidebending in the thoracic or lumbar areas of the vertebral column (for example, a scoliosis) will tend to be compensated in the cervical spine to ensure that the eyes remain in the horizontal plane. The converse may also be true. If the upper cervical spine is locked in sidebending/rotation a compensatory scoliosis is likely to develop below in order to keep the eyes level.

These discrepancies may frequently be seen in patients. Their proper evaluation and treatment are often complex and have produced a voluminous response in the literature of the manipulative therapies.

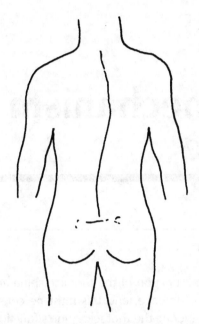

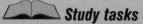 **Study tasks**

- With reference to Figure 5.13 consider the compensatory adjustments that might develop in the vertebral column as a result of leg length discrepancy (in this exercise assume a shorter left leg).
- Consider also the symptoms that might develop as a result.

Figure 5.13 Hypothetical compensatory adjustments in the vertebral column due to a shorter (left) leg, showing a thoracic and a lumbar scoliosis.

Chapter 6

The respiratory mechanism and the thoracic spine

Introduction

The structure and function of the thoracic spine are quite literally connected to the rib cage, and this must be constantly borne in mind when considering the thoracic spine's function.

📖 *Study tasks*

- Shade the components of the thoracic skeleton in separate colours (Figure 6.1 (a–b)).
- Review the features of a typical thoracic vertebra, shown in Figure 4.2 (c) (p. 25).
- Highlight the other details shown in Figure 6.1 (a–d).

(a) Thoracic skeleton (anterior view)

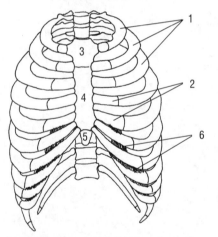

1 Ribs
2 Costal cartilage (all ribs) (hyaline)
3 Manubrium ⎫
4 Body ⎬ of sternum
5 Xiphoid process ⎭
6 Interchondral ligaments
7 Costotransverse joint
8 Costovertebral joint (joint of the head of the rib)
9 Thoracic vertebra

(b) Links between typical ribs, thoracic vertebra and sternum (horizontal section)

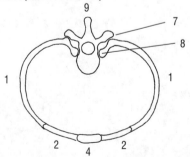

(c) Movements of the thorax during inspiration

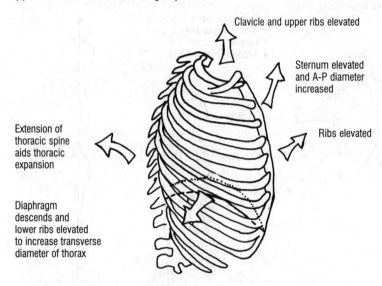

Clavicle and upper ribs elevated

Sternum elevated
and A-P diameter
increased

Extension of
thoracic spine
aids thoracic
expansion

Ribs elevated

Diaphragm
descends and
lower ribs elevated
to increase transverse
diameter of thorax

(d) Movements of the thorax during expiration

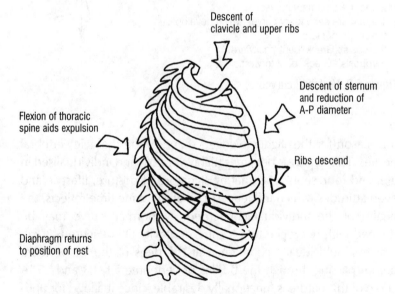

Descent of
clavicle and upper ribs

Descent of sternum
and reduction of
A-P diameter

Flexion of thoracic
spine aids expulsion

Ribs descend

Diaphragm returns
to position of rest

Figure 6.1 Selected features of the thoracic skeleton

The primary thoracic curve

The flexion curve that is apparent in the thoracic spine is referred
to as the thoracic 'kyphosis' (Greek 'bent forwards') and develops
from the primary curve that is formed *in utero* (Figure 6.2 (a)). The
apex of the curve is usually at approximately T6–8, and is probably

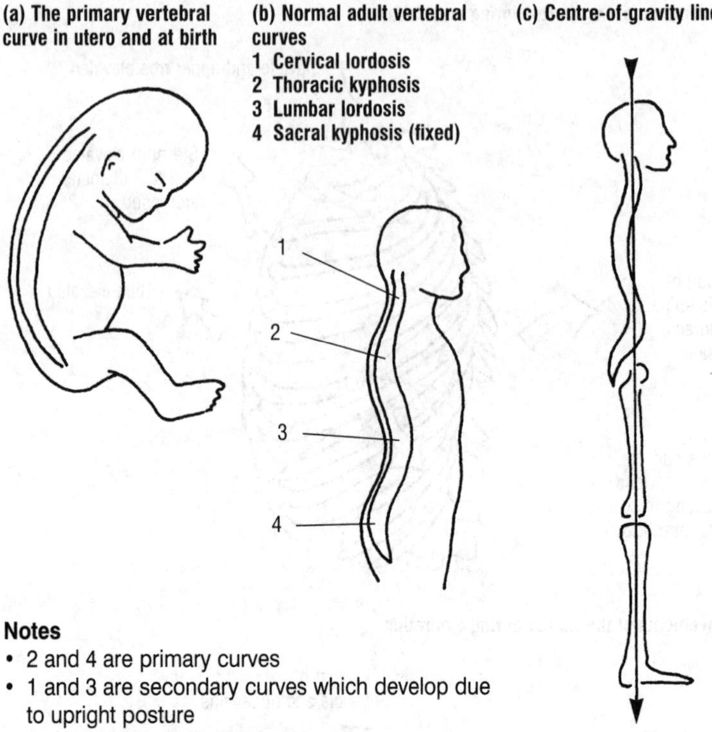

(a) The primary vertebral curve in utero and at birth

(b) Normal adult vertebral curves
1 Cervical lordosis
2 Thoracic kyphosis
3 Lumbar lordosis
4 Sacral kyphosis (fixed)

(c) Centre-of-gravity line

Notes
- 2 and 4 are primary curves
- 1 and 3 are secondary curves which develop due to upright posture
- 'Lordosis': Greek 'bent backwards'.
- 'Kyphosis': Greek 'bent forwards'.

Figure 6.2 Vertebral curves

Clinical Note

Exaggeration of the normal kyphosis will tend to adversely affect the function and health of the thoracic spine and respiratory mechanism. It will also have implications for the cervical and lumbar curves. These and related issues will be explored in the sections that follow.

---

the reason for the slightly wedged shape of the thoracic vertebral bodies. This kyphosis becomes highly variable and individualised in life, and represents a fascinating signature of posture, lifestyle and even attitudes, which influences, and at the same time reflects, the health of the individual concerned. A 'normal' curve may be defined with reference to a 'centre of gravity line' (Figure 6.2 (c)), which should ideally commence at the dens of the C2 vertebra before passing through the bodies of vertebrae T1, L3 and L5. A curve of this nature is functionally desirable, since it allows for optimal respiratory movement while at the same time interacting with the extension curves (lordoses) in the cervical and lumbar regions (Figure 6.2(b)) to aid shock absorption.

The bones

There are 12 thoracic vertebrae and ribs, whose individual structure reflects the fact that the function of the upper thoracic spine is

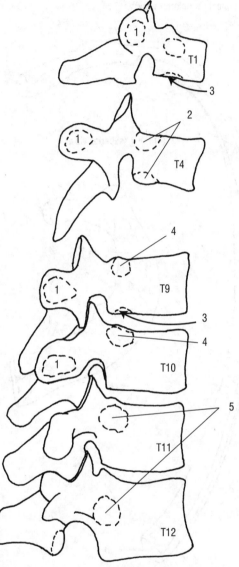

1 Costal facet for rib tubercle
2 Demi-facet for rib head
3 Small semi-lunar facet for rib head
4 Larger semicircular facet for rib head
5 Complete circular facet for rib head

 Study tasks

- Consider the following postural habits and activities, and how successfully the thoracic kyphosis might adapt to these demands: (i) sitting slumped; (ii) driving; (iii) working at a desk; (iv) jumping; (v) standing.
- Observe the effects on your cervical and lumbar spine of deliberately rounding your shoulders and flexing your thoracic spine. Do this in front of a long mirror if possible.

Figure 6.3 Features of selected thoracic vertebrae showing variations in facets for rib attachments (lateral view)

very different from the lower thoracic spine in terms of both vertebral and respiratory mechanics.

The upper vertebrae are smaller (T1 is similar in many ways to C7) but increase in size downwards; whereas the lower vertebrae are similar in size and appearance to upper lumbar vertebrae. The presence of specialised costal facets exclusively on the thoracic vertebrae is an indication of where and how the ribs attach.

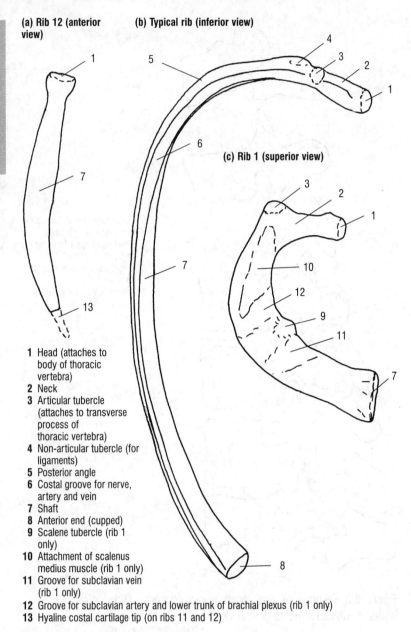

(a) Rib 12 (anterior view)

(b) Typical rib (inferior view)

(c) Rib 1 (superior view)

1 Head (attaches to body of thoracic vertebra)
2 Neck
3 Articular tubercle (attaches to transverse process of thoracic vertebra)
4 Non-articular tubercle (for ligaments)
5 Posterior angle
6 Costal groove for nerve, artery and vein
7 Shaft
8 Anterior end (cupped)
9 Scalene tubercle (rib 1 only)
10 Attachment of scalenus medius muscle (rib 1 only)
11 Groove for subclavian vein (rib 1 only)
12 Groove for subclavian artery and lower trunk of brachial plexus (rib 1 only)
13 Hyaline costal cartilage tip (on ribs 11 and 12)

Figure 6.4 Features of selected ribs

From T3 to T7 the ribs assume a fairly typical structure with a twisted shaft that allows the anterior part to connect either directly or indirectly to the sternum via hyaline costal cartilage. A groove on the inferior and posterior surface of the shaft offers protection for an intercostal nerve, artery and vein.

Rib 1 is the shortest and most curved rib, with a tubercle on its upper surface where the scalenus anterior muscle attaches. The subclavian vein and artery rest on shallow grooves on either side of this tubercle. The scalenus medius muscle is also attached to the upper surface.

Rib 2 (not shown) has a similar shape but is almost twice as long. It gives attachment to the scalenus posterior muscle on its upper surface, and so the three scalene muscles elevate the upper two ribs in inspiration.

The upper seven ribs attach anteriorly and directly to the sternum by means of hyaline costal cartilage. These are the so-called 'true ribs'. The remaining ribs either 'join on' to the costal cartilage above (ribs 8–10) or 'float' without attachment to the sternum or transverse processes of the vertebrae. Ribs 11 and 12 are almost dagger-like and tipped with costal cartilage. Ribs 8–12 are sometimes called by the rather misleading term 'false ribs', since they do not attach directly to the sternum. Vertebrae T9–12 reflect these rib changes through a change in the location of costovertebral facets, and, in the case of T11 and T12, the absence of costotransverse facets (Figure 6.3).

The sternum

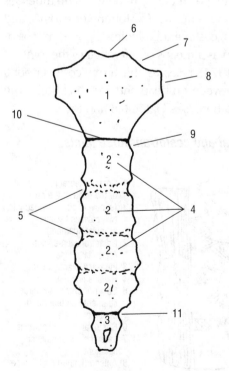

1 Manubrium sterni
2 Body (mesosternum)
3 Xiphoid process (may be perforated)
4 Sternebrae (segments)
5 Facets for costal cartilage
6 Jugular notch
7 Facet for clavicle
8 Facet for costal cartilage of rib 1
9 Facet for costal cartilage of rib 2
10 Manubriosternal joint (symphysis)
11 Xiphisternal joint (symphysis)

Figure 6.5 The sternum (anterior view)

Study tasks

- Shade the three bones of the sternum and highlight the details shown.
- Palpate the following structures on yourself: (i) jugular notch; (ii) sternoclavicular joint (p.100); (iii) first sternocostal joint; (iv) manubriosternal joint; (v) sternal angle; (vi) rib 2; (vii) xiphoid process; (viii) xiphisternal joint.
- Practise counting ribs on yourself and a colleague. Start close to the sternum, but also palpate posteriorly to find the lowest ribs.

The sternum (Greek 'chest') consists of an arrangement of three bones which resembles a short sword, or dagger. Hence, the superior part is called the 'manubrium' (Latin 'handle'), which is linked to the 'body' (middle bone) by a symphysis forming the manubriosternal joint. The body or 'mesosternum' develops from four sternebrae (segments) which fuse, and it has notches at its edges for the attachment of the rib costal cartilages. Inferiorly, the sternum is completed by a tip of bone known as the xiphoid process, or 'xiphisternum' (Greek *xiphos* = 'sword'), linked to the sternal body by a symphysis, the xiphisternal joint. These structures are easily palpable, and a slight change of direction of the manubrium on the mesosternum is palpable at the manubriosternal joint and referred to as the sternal angle (of Louis). The rib 2 attaches at this point, which forms a useful palpatory landmark. Above the manubrium lies the jugular (suprasternal) notch. The first seven costal cartilages join to the sternum on each side.

The joints and ligamentous structures

Note: the terminology of the posterior costovertebral joints varies. Williams (1995) refers to both the articulations between rib tubercles and transverse processes of the vertebrae (costotransverse joints) and the joints of the rib heads as 'costovertebral joints'. The term 'costo-corporeal' is also introduced as a means of reference to the joints of the rib heads. Kapandji (1974) reserves the term 'costovertebral joint' for the articulation between rib head and vertebral body. The latter convention is adopted here (see Figure 6.6 and text).

The posterior costovertebral and costotransverse joints

(a) Anterolateral view (right)

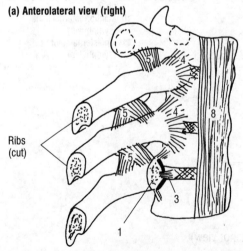

Ribs
(cut)

1 Costovertebral joint – joint of the head of the rib (synovial, plane divided into two cavities by intra-articular ligament)
2 Costotransverse joint (synovial, plane)
3 Intra-articular ligament
4 Radiate ligament
5 Superior costotransverse ligaments (anterior and posterior layers)
6 Costotransverse ligament
7 Lateral costotransverse ligament
8 Anterior longitudinal ligament

(b) Superior view

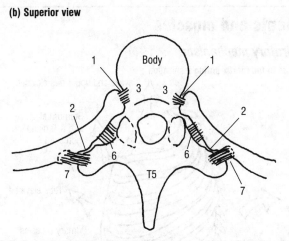

> **Study task**
> • Shade the ligaments in colour and highlight all features shown in Figures 6.6 and 6.7.

Figure 6.6 Costovertebral and costotransverse joints with associated ligaments

The anterior sternocostal and interchondral joints

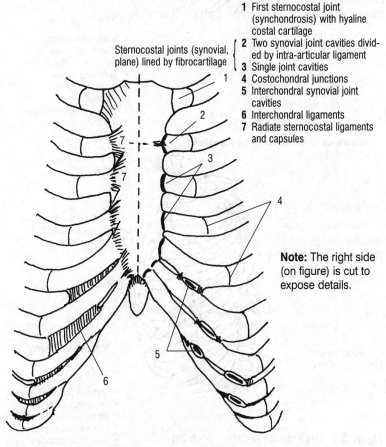

Sternocostal joints (synovial, plane) lined by fibrocartilage

1 First sternocostal joint (synchondrosis) with hyaline costal cartilage
2 Two synovial joint cavities divided by intra-articular ligament
3 Single joint cavities
4 Costochondral junctions
5 Interchondral synovial joint cavities
6 Interchondral ligaments
7 Radiate sternocostal ligaments and capsules

Note: The right side (on figure) is cut to expose details.

Figure 6.7 Sternocostal and interchondral joints (anterior view)

Movements and muscles

The respiratory mechanism

(a) Movements of the thorax during inspiration

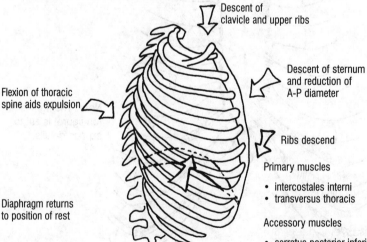

Clavicle and upper ribs elevated

Sternum elevated
and A-P diameter
increased

Ribs elevated

Extension of
thoracic spine
aids thoracic
expansion

Diaphragm
descends and
lower ribs elevated
to increase transverse
diameter of thorax

Primary muscles

- diaphragm
- levatores costarum
- intercostales externi
- scalenus anterior
- scalenus medius
- scalenus posterior

Accessory muscles

- iliocostalis cervicis
- sternocleidomastoid
- pectoralis major
- pectoralis minor
- latissimus dorsi
- serratus anterior
- serratus posterior superior
- quadratus lumborum

(b) Movements of the thorax during expiration

Descent of
clavicle and upper ribs

Descent of sternum
and reduction of
A-P diameter

Flexion of thoracic
spine aids expulsion

Ribs descend

Diaphragm returns
to position of rest

Primary muscles

- intercostales interni
- transversus thoracis

Accessory muscles

- serratus posterior inferior
- latissimus dorsi
- iliocostalis lumborum
- longissimus thoracis
- abdominal muscles
- subcostales

Figure 6.8 The thoracic skeleton, showing
respiratory movements (anterolateral view)

As a result of the configuration of ribs, vertebrae and ligamentous structures, the thoracic skeleton is a dynamic structure which expands during inspiration (breathing in) due to the descent of the diaphragm and contraction of the other muscles of inspiration, which allows the lungs to fill with air. Expiration (breathing out) is a reversal of this process, involving natural tissue recoil, the force of gravity, and certain muscles of expiration. The degree of muscular activity depends on whether breathing is 'quiet' (shallow) or 'forced' (deep), in either inspiration or expiration.

The muscles that activate quiet breathing are used at all times and are called 'primary' (principal) muscles of inspiration and expiration. The muscles that are recruited for the additional efforts of deeper breathing are called accessory muscles of inspiration and expiration (Figure 6.8).

The structure of the thoracic skeleton is also designed to protect internal organs and to allow efficient movement, as well as functioning as a respiratory mechanism. Therefore, the upper part of the thoracic cage is relatively enclosed to protect the lung apices from the vigorous mobility of the shoulder joints. Upper thoracic spinal movements are also restricted by the protective rigidity of the upper rib cage.

In contrast, the lower ribs are somewhat flared, mobile and open, since the lumbar vertebral column demands greater flexibility.

Rib movements in respiration

It is not surprising to find that there are differences between the motion of the upper and lower ribs during respiration (Figure 6.9). In the upper two or three ribs the axis of movement between the costovertebral and costotransverse joints is relatively coronal, and this encourages the almost ring-shaped thoracic cage to move anteriorly and superiorly aided by concave costal facets on the transverse processes of the vertebrae (Williams, 1995). This movement has been likened to the action of a pump handle (Snell, 1995) (Figure 6.9(a)). It could also be likened to the rise and fall of a ring-shaped door knocker.

In the lower ribs there is a shift in the axis of movement towards the sagittal plane, which, together with a flattening of the costotransverse facet surfaces, encourages the lower ribs to flare outwards: this increases the transverse diameter of the thoracic cage. The latter movement has been likened to the action of a bucket handle (Kapandji, 1974; Snell, 1995). The ribs in-between are transitional and appear to blend these movements.

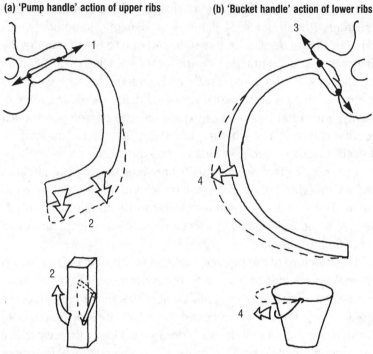

(a) 'Pump handle' action of upper ribs

(b) 'Bucket handle' action of lower ribs

1 Semicoronal axis of movement
2 'Pump handle' action of upper ribs increasing A-P diameter of thorax
3 Semisagittal axis of movement
4 'Bucket handle' action of lower ribs increasing transverse diameter of thorax

Figure 6.9 Rib movements in respiration

Both the sternum and vertebral column must accommodate these movements and this can be summarised diagrammatically:

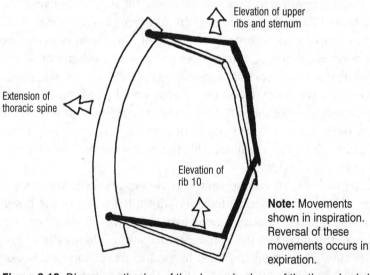

Elevation of upper ribs and sternum

Extension of thoracic spine

Elevation of rib 10

Note: Movements shown in inspiration. Reversal of these movements occurs in expiration.

Figure 6.10 Diagrammatic view of the change in shape of the thoracic skeleton during respiration (sagittal plane)

📖 **Study tasks**

- Palpate your own supraclavicular fossa with one hand (either side will do) while placing the other hand over the lower ribs laterally. Gentle but firm downward pressure in the supraclavicular fossa will be met by resistance from the first rib. Take a deep breath and you should be able to feel the contraction of the scalene and sternocleidomastoid muscles which elevate the upper two ribs and sternum in inspiration. Notice that this is an upward and forward movement, whereas in contrast the lower ribs move outwards. You should be able to palpate the difference between the 'pump handle' and 'bucket handle' movements of the upper and lower ribs.
- Observe and palpate these movements in a colleague. You will find that this is more effective if you stand behind your colleague and use both hands to palpate on each side.
- Muscle check (see Figure 6.8).

It is not easy to observe and measure breathing patterns in individuals (including oneself) because the rate and intensity of breathing may be consciously altered. Try the study tasks on the right on yourself as well.

Study tasks

- Observe a colleague quietly breathing in both upright and supine positions, noting any differences.
- Ask your colleague to take a few deep breaths, and note any differences.
- Ask your colleague to exhale forcibly and also to cough. Try to identify the muscles that are being used.

Clinical Note

Breathing is such a fundamental aspect of health that observation of the shape and development of the thoracic skeleton may reveal signs of respiratory disorders. For example, the thoracic skeleton of the asthma and bronchitis sufferer may show signs of enlargement or change in shape. Excessive use of the upper ribs and hypertrophy of the associated muscles may also be apparent. Varying rates of breathing and breathing patterns have always been important signs in clinical medicine. It is beyond the scope of this book to explore these issues further, but it is worth noting that the student who wishes to study respiratory health should first be thoroughly familiar with the practical anatomy of the thoracic skeleton.

The thoracic spine: movements

The structural support and protection offered by the thoracic vertebrae and ribs result in a degree of rigidity which inhibits movement in the thoracic spine compared with elsewhere. The proximity of the more mobile cervical and lumbar regions means that there are also considerable functional implications for the cervicothoracic and thoracolumbar junctional areas, which will be examined separately in Chapter 8.

The vertebral bodies are slightly wedged in shape (deeper posteriorly) and the intervertebral discs are relatively smaller than in either the cervical or lumbar regions, with a disc:vertebra ratio of 1 : 5 (Kapandji, 1974). These two factors give rise to the flexed thoracic kyphosis, which is a reminder of the original primary flexion curve of the spine. This flexion curve contributes to the protective shape of the thoracic cage, but may also confer an unwelcome degree of rigidity to the spine in functional terms. The kyphosis also alters the functional plane of the apophyseal joints.

Clinical Note

Exaggeration of the thoracic kyphosis will result in a compensatory exaggeration of the cervical lordosis above. This, together with a tendency for the upper thoracic apophyseal joints to become weightbearing and static, results in much discomfort in predominantly sedentary societies due to functional loss, physiological stasis and muscular hypertonia.

 Study tasks

- Highlight the details shown in Figure 6.12.
- Muscle check.
- Consider the factors that limit the movements of flexion and extension in the thoracic spine.
- Observe variations in the degree of kyphosis between colleagues and in a wider population. Consider such influences as age, lifestyle and types of employment on this part of the vertebral column. Consider any other factors that might be relevant.
- With a colleague seated with hands clasped behind their neck to reduce movements in the cervical spine, support them safely and rock them backwards and forwards gently, using one hand to palpate degrees of flexion/extension and the other to grip their elbows for leverage.

Figure 6.11 Changes in the orientation of apophyseal joint surfaces in the thoracic spine due to the thoracic kyphosis (sagittal view)

Flexion/extension

(a) Flexion

Movement produced by:
- mostly indirect action via the abdominal muscles and hip flexors
- assisted by intercostal muscles
- aided by gravity

(b) Extension

Movement produced by:
- erector spinae
- transversospinales
- intertransversarii
- interspinales
- assisted by middle/lower fibres of the trapezius

Figure 6.12 Flexion/extension of the thoracic spine (detail of motion segment)

As in other vertebral areas, the movement of flexion results in the inferior articular processes sliding upwards and forwards on the superior articular processes of the vertebra below. The anterior structures including the intervertebral disc are compressed, while the posterior structures and ligaments are stretched. The nucleus pulposus is squeezed posteriorly. In extension, the process occurs in reverse.

As the thoracic kyphosis increases within the individual spine so the range of available movement decreases. Thus at T1–6 flexion/extension may be as little as 4°, rising to 6° at T7–10, and 9° at T11–L1, reflecting the proximity of the lumbar spine. Within this range as much as 60–70% has been attributed to flexion, with extension contributing only 30–42% (adapted from White and Panjabi, 1978a).

Sidebending (lateral flexion)

Sidebending is limited in the thoracic spine by the presence of the ribs, but increases progressively down towards the more mobile lumbar area. Ranges of 6° movement at T1–9 rising to 8° at T10–12 have been suggested (adapted from White and Panjabi, 1978a). Sidebending is always accompanied by an element of rotation (see study task below).

The following experiment shows, however, that in certain positions, the rotation that accompanies sidebending in the mid-thoracic spine may occur to the opposite side.

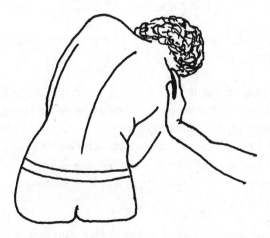

Figure 6.13 Passive sidebending of the thoracic spine in flexion

 Study task

• Demonstrate sidebending with rotation using a flexible plastic spine with a normal (not exaggerated) thoracic kyphosis. You should find that the apophyseal joint surfaces control the direction of movement. Theoretically, when the vertebral column sidebends, the articular facets on the same side are compressed and the inferior articular process slides down on the sloping surface of the superior articular process of the vertebra below. This produces rotation with posterior translation to the same side. This process has already been observed in the cervical spine, where the oblique angle of the apophyseal articular surfaces ensures that they control all cervical movements.

Study tasks

- Seat a colleague in a flexed position (ask them to slump) and support them with your hand as they lean to the right, as illustrated in Figure 6.13. It is important that they use as little active muscle control as possible. You should be able to see the concavity of the sidebending curve on the right, but the ribs will normally be more prominent on the left (the so-called 'high side'). This indicates that rotation of the vertebrae has taken place to the opposite side of the sidebending movement.
- Repeat the experiment but this time with the same colleague sitting in slight extension. This is an attempt to shift the centre of gravity posteriorly and should produce a degree of rotation towards the concavity (with the high side now on the right). Try not to overdo the amount of extension or the thoracic spine will in effect lock and any discernible movement will come from the lumbar rather than the thoracic spine.

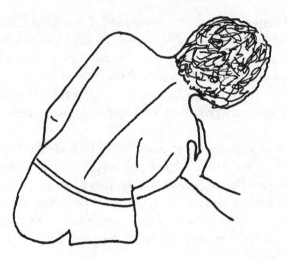

Figure 6.14 Passive sidebending of the thoracic spine in slight extension

The reason for this apparent paradox may have been first investigated by Lovett (1900; 1902). He separated the anterior part of the motion segments from the posterior part (see Figure 4.2(a) p. 25) and then noted that the two halves behaved differently in sidebending and rotation. It was Fryette (1954), however, who made the following observation:

> When the spine is in a position that the (apophyseal) facets are not functioning: when they are 'idling', or in other words when they are in neutral, the spine takes on the characteristics of a pile of blocks, and when it is sidebent, it tends to collapse toward the convexity, the bodies crawl out from under the load.

An experimental model of this is described by Kapandji (1974). Fryette (1954) also observed,

> To the degree that the facets are in control in any area of the spine, they direct and govern rotation. Therefore, as facet control increases, the spine takes on the characteristics of a flexible ruler or a blade of grass – it must be rotated before it can be sidebent – and when it is rotated and sidebent the bodies are forced to the concavity of the curve produced by this sidebending'.

If we are to take these statements at face value we are led to the following conclusions:

1. Whenever the vertebral column sidebends and exhibits rotation towards the concavity, the posterior parts of the motion segments (the apophyseal joints), are controlling the movement.
2. Whenever the vertebral column sidebends and exhibits rotation towards the convexity, the anterior parts of the motion segments (the intervertebral discs and bodies) are controlling the movement.

In the lower cervical spine (C3–7), sidebending and rotation are always to the concavity in a normal spine because the apophyseal joints always control movements due to the 45° inclination (approximately) of the apophyseal joint surfaces. In the thoracic spine the situation is complicated by the following factors:

- the varying angles of apophyseal joint surfaces, which are mainly in the coronal plane but subject to variations through the thoracic spine (see Figure 6.11);
- the slightly wedged shape of the vertebral bodies, which helps to form the kyphosis (which varies between individuals);
- the presence of the ribs and their supporting ligaments;
- the possible presence of anomalies such as lateral curvature (scoliosis), osteochondrosis, or asymmetrical muscle contraction.
- it is difficult, if not impossible, to isolate the lumbar movements coming from below.

These complicating factors mean that if the mid-thoracic spine is placed in varying degrees of flexion/extension, and then sidebent, the direction of rotation may vary. Also note that the thoracic spine may not rotate at all if the thoracic region locks in full extension. In the latter case the main movement may be observed coming from the lumbar spine below.

Rotation

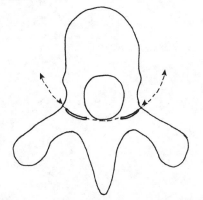

Movement produced by:

(on the same side)
- obliquus internus abdominis
- levatores costarum

(on the opposite side)
- rotatores
- multifidus
- semi spinalis thoracis
- obliquus externus abdominis

Figure 6.15 The orientation of thoracic apophyseal joint surfaces facilitating rotation movements (superior view)

 Study task

- Consider reasons for the differences in rotation patterns between a functional and a structural scoliosis. Also consider the possibility of a transitional state.
- Muscle check.

Rotation in the thoracic spine is encouraged by the plane and orientation of the apophyseal joint surfaces, especially in the upper/middle regions. An element of sidebending also accompanies the movement.

Study task

• Seat a colleague in an upright position on a stool with their hands clasped behind their neck in order to protract the scapulae and expose their thoracic spine. Grasp their elbows and gently rotate them to the right. Notice the curvature through the thoracic spine, with the concavity and high side on the right. Rotation and sidebending are both to the same side, since the apophyseal joints are controlling the movement. Notice also that the degree of rotation seems to decrease down the thoracic spine, which is in accordance with the findings of White and Panjabi (1978a) of 9° rotation at T1–2, 8° at T2–8 and 2–7° at T8–L1.

Figure 6.16 Passive rotation of the thoracic spine

Chapter 7

The lumbar spine

Introduction

The lumbar spine is the main load-bearing section of the vertebral column in human beings. The peculiarly upright stance, coupled with the demands of domestic and economic activity, means that the human being is the only creature who, rather unwisely, lifts heavy weights, sits for prolonged periods of time, drives vehicles and, as a result, commonly suffers back pain.

It is fortunate that the structure of the lumbar spine reflects the greater demands placed upon it, but equally unfortunate that few people understand the functional limits within which they should be operating. It is the author's opinion that simple principles of back care should be taught not only in the workplace but also in schools.

Note that the functional unit referred to as 'the low back' should be properly defined as a wider area than just the lumbar spine, to include the sacro-iliac joints, as well as the fascia and muscles attaching to the thoracolumbar and hip regions, and beyond. Thus the term 'lumbar spine' as discussed in this chapter should not be confused with the concept 'low back', even though many of the observations may be relevant to both.

The bones

There are five lumbar vertebrae, whose structure progressively reflects the change from thoracic to lumbar weightbearing functions. This means that it is difficult to isolate any individual lumbar vertebra as 'typical'.

- Highlight the features shown.
- Study the differences between individual lumbar vertebrae using either bone specimens or a flexible plastic spine.
- Try to discover the functions of the mamillary and accessory processes.
- Compare the L5 vertebra with an upper lumbar vertebra; and with reference to Figure 7.2(b–c) consider why the transverse processes of L5 are more massive.

(a) Selected features of the lumbar spine

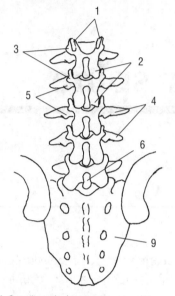

(b) The five lumbar vertebrae (lateral view)

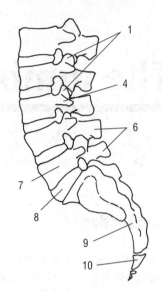

1 Superior articular process
2 Inferior articular process
3 Mamillary process
4 Transverse process (longest at L3)
5 Accessory process (at root of transverse process)
6 Spinous process
7 Vertebral body (L5 is wedge shaped)
8 Intervertebral disc (all lumbar discs are wedge shaped)
9 Sacrum
10 Coccyx

Figure 7.1 Selected features of the lumbar spine

The joints and ligamentous structures

Chapter 4 introduced the concept of the vertebral 'motion segment' (functional spinal unit), consisting of two adjacent vertebrae with intervening discs and ligaments, but excluding the surrounding muscles and fascia. The motion segment is subdivided into anterior and posterior parts.

(a) Ligaments and main features of a lumbar vertebral motion segment (lateral view)

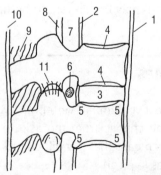

(b) The ilio-lumbar ligament (posterior view)

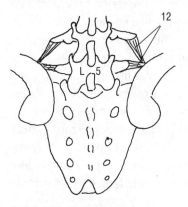

(c) The ilio-lumbar and lumbosacral ligaments (anterior view)

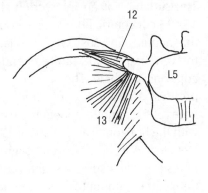

  1 Anterior longitudinal ligament
  2 Posterior longitudinal ligament
  3 Intervertebral disc
  4 Hyaline cartilage vertebral end plate
  5 Ring epiphyses (growth zones in immature bone)
  6 Nerve root in intervertebral foramen
  7 Vertebral canal (spinal cord)
  8 Ligamentum flavum
  9 Interspinous ligament
 10 Supraspinous ligament
 11 Apophyseal joint (with enclosing fibrous capsule)
 12 Ilio-lumbar ligament (from transverse process of L5 (and sometimes L4) to iliac crest)
 13 Lumbosacral ligament (anterior lower band of ilio-lumbar ligament)

Figure 7.2 Ligaments of the lumbar spine

The anterior part of the motion segment

The vertebral bodies

The trabecular structure of the vertebral bodies is designed to bear compressive loads, although in elderly or osteoporotic vertebrae the surrounding cortical bone shell may contribute as much as 75% of the support (Rockoff *et al.*, 1969; Yoganandan *et al.*, 1988).

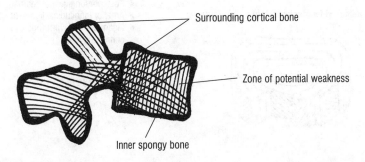

Surrounding cortical bone

Zone of potential weakness

Inner spongy bone

Figure 7.3 Schematic view of internal trabeculae in a typical lumbar vertebra (sagittal section)

Clinical Note

The internal cancellous or spongy bone of the vertebral body has spaces containing red marrow for red blood cell production. It is, however, a zone of potential mechanical weakness (Figure 7.3), which may produce abnormal 'wedging' in the case of developmental softening of the vertebral bodies, known as osteochondrosis; and in later life may result in osteoporotic 'crush' fractures.

The intervertebral discs

The structure of an intervertebral disc consists of approximately 15–25 concentric fibrocartilaginous lamellae known collectively as the 'annulus fibrosus', whose collagen fibres are firmly anchored to the vertebral hyaline cartilage 'endplates' and ring epiphyses of the adjacent vertebrae. The fibres of the annulus fibrosus surround and prestress a gelatinous core (the 'nucleus pulposus'). The collagen fibres of the annulus appear to be arranged at an angle of about 30° with the horizontal and at about 120° to each other (Figure 7.4(b)). Detailed observation reveals slight variations in angle, so that the fibres actually assume a slightly curved course, and there may also be a certain amount of merging between the lamellae (Marchand and Ahmed, 1990).

Study task

• Shade the nucleus pulposus, annulus fibrosus, and ligaments in separate colours in Figure 7.4.

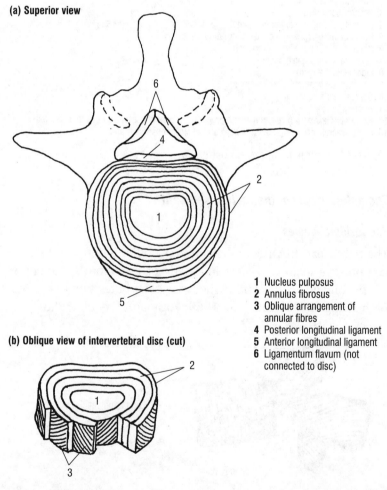

(a) Superior view

(b) Oblique view of intervertebral disc (cut)

1 Nucleus pulposus
2 Annulus fibrosus
3 Oblique arrangement of annular fibres
4 Posterior longitudinal ligament
5 Anterior longitudinal ligament
6 Ligamentum flavum (not connected to disc)

Figure 7.4 Schematic views of the structure of an intervertebral disc with supporting ligaments

The nucleus pulposus consists of a hydrophilic (water-binding) matrix of proteoglycan water-gel (glycosaminoglycans), and a loose mix of collagen fibres.

The disc tends to lose its water content during compressive loading, so that during a typical day body height may be reduced by as much as 20 mm (De Pukey, 1935; Tyrell *et al.*, 1985; Krag *et al.*, 1990). This is replaced during rest periods (i.e. when not weight-bearing) by 'imbibation' of tissue fluid from the adjacent vertebral bodies. There is no direct blood supply to the discs. The proteoglycan content diminishes with age and/or degeneration, so that the disc tends to become progressively less hydrated.

The nucleus is approximately central in cervical and thoracic discs, but lies in a slightly posterior position in the lumbar spine. It is under hydrostatic pressure (equal in all directions), and the main function of the annulus is to contain this. Measurement has shown an intrinsic pressure of approximately 10 N/cm^2 in unloaded discs within the nucleus due to forces exerted by the annulus, the ligamenta flava and the longitudinal ligaments.

Loading of the lumbar spine exerts a compressive force which is derived from the weight of the upper body and the tensile force of the vertebral muscles and fascia. Compressive stress in the nucleus is about 1.5 times the external load, compared with 0.5 times the load in the annulus. The annulus bears predominantly tensile stress, which can be 4–5 times the external load per unit area in the posterior annulus (Nachemson, 1960), and this is particularly important in bending and rotational movements. The nucleus withstands compression by converting the vertical stress into circumferential tensile and radial compressive stress in the annulus (Nachemson, 1960). This causes the annulus to bulge radially outwards and also into the vertebral endplates between the adjacent vertebrae.

(a) Light loading   **(b) Heavy loading**

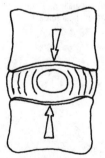 

Increasing compression causes annulus to bulge and raises pressure on vertebral end plates

Figure 7.5 Schematic view of compressive loading of vertebral motion segment

Study tasks

- Highlight the features shown in Figure 7.5.
- Simulate the loading of a disc by pressing hard on a squash ball (or similar) on a table. The pressure you feel against your hand simulates the forces acting on the vertebral endplate. Note also the radial bulging.

Note: Further types of injury to the intervertebral disc will be discussed under 'flexion' below.

The posterior part of the motion segment

It has been consistently emphasised that the main role of the vertebral articular processes (and hence the apophyseal facet joints) is to guide movements. Although all three axes of movement are possible throughout the vertebral column, the actual range of available movement at cervical, thoracic and lumbar levels is largely determined by the configuration of the articular processes.

The orientation of the apophyseal joint surfaces of vertebrae L1–4 is basically sagittal, but approximately at a 45° angle to the frontal plane. This allows flexion/extension mainly, and a degree of sidebending, but only limited rotation. This orientation becomes progressively more coronal down the lumbar spine in order to counteract forward shear forces.

The presence of the lumbar lordosis means that the lower lumbar vertebrae are inclined at a considerable angle to the horizontal, which reaches a maximum at the lumbosacral junction between L5 and the sacrum, and forms the 'lumbosacral angle' (Figure 7.6) as defined by Cailliet (1981; 1995) but not Kapandji (1974), who refers to this angle as the 'angle of the sacrum'. This angle results in a gravitational shear force acting to move each vertebra forwards over the one below.

The lumbosacral angle

1 Lumbosacral angle (angle of the sacrum) – usually *c*. 30–50°
2 Direction of shear force

Figure 7.6 The lumbosacral angle

If the vertebrae and discs are loaded with combined compression and shear, as is usual in life, the neural arch (which includes the articular processes) is said to resist about 70% of the shear component, with the disc resisting the rest (Adams, 1994).

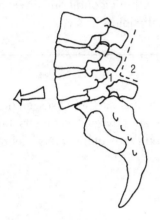

1 Defect in pars interarticularis at L5 has led to anterior displacement of vertebral column, or 'spondylolisthesis'
2 Palpable 'shelf sign'

Figure 7.7 Spondylolisthesis (lateral view)

> ### Clinical Note
>
> *A defect in the lamina between the superior and inferior articular processes (the 'pars interarticularis') can lead to a hairline fracture known as a 'spondylolysis'. The shear forces present in the lumbar spine due to the lordosis can lead to a forward slippage which is known as a 'spondylolisthesis'.*

This defect can occur at other levels, but is more likely to develop in the lower lumbar spine. The effects are greatly minimised by the supporting strength of the iliolumbar ligament, the orientation of the articular facets at L5, and the surrounding muscles.

The apophyseal facet joints and inclusions

These are generally regarded as plane synovial joints, which guide movements in the lumbar spine as elsewhere. The details of these movements are discussed below. However, their load-bearing role should also be considered, especially in view of the lumbar lordosis and the quite common postural exaggeration of this feature.

The apophyseal facet joints are quite complex in structure. They have a fibrous capsule enclosing the joint surfaces, lined by synovial membrane, which is sometimes shaped into folds. The capsule must be lax enough to allow movement, but tight and strong enough to give stability. Perhaps in order to facilitate this, a variable vascular fatty layer may be present between the capsule and synovial surfaces, which could allow the synovial membrane to fill potential gaps around the articular margins of the joint (Giles and Taylor, 1987).

Santo (1935) may have been one of the first to describe the

Clinical Note

The presence of inclusions within the apophyseal joints, whether they be menisci, fatty synovial folds or even small pieces of detached articular cartilage, may explain why these joints sometimes painfully lock, yet can apparently be released by manipulation.

Study tasks

- Consider the effect of the removal of the nucleus pulposus, that is, loss of disc height, on the articular processes and facet joints of the lumbar spine.

presence of fibrous inclusions with cartilage cells which would seem to be true menisci. These findings are also supported by Lewin *et al.* (1961).

An excessive degree of weightbearing imposed on these joints in the lumbar spine is a potent cause of early degenerative changes, and may give rise to the degenerative arthritic condition known as 'spondylarthrosis' (literally 'arthritis of the joints').

Movements and muscles

The lumbar spine permits movement through all three axes, guided and stabilised by the orientation of the apophyseal facet joints, but probably influenced also by the wedged shape of the intervertebral discs and the vertebral body of L5.

Flexion

Movement produced by:
- gravity
- eccentric contraction of posterior vertebral muscles

(from standing)

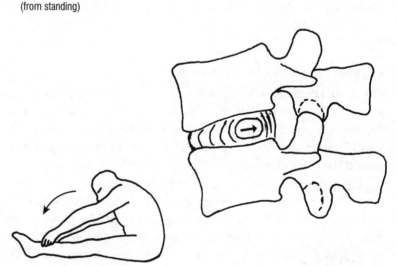

- abdominal muscles
- psoas minor
- hip flexor muscles
(from sitting)

Figure 7.8 Flexion of the lumbar spine (detail of motion segment)

Flexion brings the anterior structures closer together, increases the pressure on the anterior part of the disc and stretches the posterior structures. Intradiscal pressure has been shown to increase by up to 80% in full flexion, and the performance of the disc appears to vary with its fluid content, showing diurnal variations (Adams, 1994). Both early in the day and early in Life a well endowed nucleus seems more vulnerable to rupture. These effects will be discussed below. Some of the ligaments, such as the ligamentum flavum, appear to possess stretch receptors which probably inhibit movement at the end of range. In health, the amount of flexion seen at each motion segment is about 8° at L1–2, and 11° at L4–5 and L5–S1 (Adams and Hutton, 1982; Pearcy and Tibrewal, 1984), although movement may be expected to diminish with age at L5–S1 due to the effects of degenerative changes (see below). Flexion (*c.* 55°) exceeds extension (*c.* 30°) in the lumbar spine, but the act of lumbar flexion (as in bending down to pick something up) also involves a considerable degree of hip flexion and anterior rotation of the pelvis, which is limited to varying degrees by the hip extensor muscles. This is something that should be carefully noted in patients who may be 'protecting' their lumbar spine by predominantly using hip flexion.

The problem of lifting

On completion of the final Study Task on the right you have just begun to consider the important problem of compressive loading on the vertebral column. To simplify matters, this movement will be considered initially in terms of flexion/extension, but it should be remembered that in Life rotation forces (torque) are likely to complicate matters further.

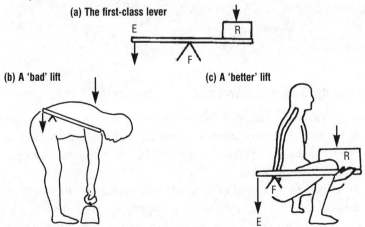

Figure 7.9 The problem of lifting

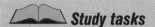 **Study tasks**

- Highlight the details shown in Figure 7.8.
- Muscle check.
- Consider the tissues that limit flexion in the lumbar spine.
- Observe active flexion in a colleague by asking them to bend forwards to touch their toes. Notice the degree of hip and pelvic rotation involved. Emphasise the importance of this by asking them to imitate a patient with a stiff back. They should try to maintain their lumbar lordosis as they bend forwards.
- Repeat the movement, but now try to eliminate the hip and pelvic components by asking them to contract their abdominal muscles and curl up from above downwards. Place your hands on their pelvis and restrain them as soon as you palpate hip flexion. Notice the relatively limited amount of lumbar flexion available.
- Practise palpation by placing your colleague in a sidelying position on a treatment table. Using their flexed hips and knees as levers resting against your body, palpate the movement at each segment (this can be combined with extension). Try to be accurate about which motion segment you are palpating, and try to decide which segments are moving most and least.
- Consider the following everyday activities on the posterior vertebral structures (including muscle, fascia and skin): (i) sitting on the floor; (ii) squatting; (iii) putting on a sock; (iv) sitting in a car (this is greatly influenced by the type of car); (v) gardening.

Study tasks

- With reference to the model of the first-class lever, consider the influence of the following variables: (i) the strength of the back muscles; (ii) obesity; (iii) pregnancy.
- Consider the difficulties encountered in lifting situations in some occupations, e.g. nurses and carers. If we accept that an erect posture is less stressful to the lumbar intervertebral discs than a flexed posture, consider approaches to seating and lifting which might take advantage of this.
- Consider why therapists often recommend swimming as a method of rehabilitation for back sufferers.
- Consider the advantages and disadvantages of using a corset for either cosmetic or protective reasons.

In lifting, the compressive force on the discs produced by the contraction of the back muscles is very great. In a simple model of lifting (Figure 7.9), the vertebral column pivots about the centre of the intervertbral discs like a see-saw, or first-class lever.

The effort arm of the back muscles is relatively short (approximately 7–8 cm) and fixed. However, the resistance arm of the weight being lifted is highly variable (say 30–50 cm). In terms of simple 'moment arm analysis', the tensile force in the muscles is about 5–6 times greater than the weight being lifted. In this model, the length of the resistance arm always exceeds that of the effort arm, which results in a 'disadvantaged' lever.

Holding a weight as close as possible to an upright body minimises the disadvantage. The whole lever mechanism may also be elevated by using one's legs.

Clinical Note

One method of improving vertebral control when lifting weights is to raise intra-abdominal pressure by taking a deep breath and 'holding' it in by contracting the abdominal muscles. This, in effect, produces a muscular corset around the vertebral column which not only protects the back when lifting, but can also be employed by patients with acute back pain who may not be aware of the value of this anterior support. Cailliet (1995) suggests that the compressive effects of disc loading are significantly reduced in this manner. The effect may also be reinforced by the use of a broad belt, which can reduce the compressive force on the lumbar spine by as much as 1000 N (McGill et al. 1990).

The damaged intervertebral disc

Flexion increases the hydrostatic pressure in the nucleus and stretches the posterior annulus. Pressure is further increased by compressive forces applied to the motion segment by muscular activity or lifting weights.

Excessive compressive loading will tend to rupture the vertebral endplates; but a combination of compression and flexion may cause the nucleus pulposus to burst through the posterior annulus. Numerous experiments have been performed *in vitro* (Adams

1994) simulating both a single severe load and numerous less severe cycles (fatigue loading). The damage caused to discs increases in severity from the development of fissures in the annulus into which nuclear material leaks, to a weakening and permanent bulging of the annulus with loss of height (herniation), and ultimately to the bursting of the disc with the extrusion of nuclear material (prolapse). A detached fragment of nucleus may be referred to as a 'sequestration'.

(a) Compressive damage to vertebral end plates (ruptures that may appear on radiographs as Schmorl's nodes/nodules)

(b) Flexion damage to intervertebral discs resulting in herniation or prolapse (shown) usually in a posterolateral direction

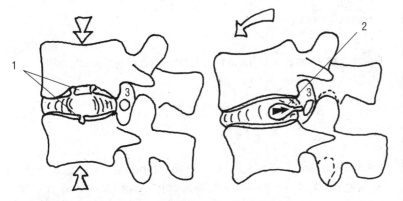

1 Fracture/rupture of vertebral endplates (Schmorl's nodes/nodules)
2 Posterolateral prolapse of nucleus pulposus and disc, compressing nerve root
3 Spinal nerve root

Figure 7.10 Two types of damage to intervertebral discs (lateral view)

Although these experiments cannot be reproduced *in vivo* there is clear evidence from epidemiological surveys that the occurrence of discal damage is very similar to that described above, and is associated with flexion and lifting (Kelsey *et al.*, 1984). The problem of disc injury is a huge topic, which cannot be fully addressed here, and the interested reader should consult the very wide literature on the subject.

Extension

Extension approximates the lumbar vertebrae posteriorly, with resistance coming from the facet joints and the apposition of the spinous processes. The posterior part of the annulus bulges outwards and experiences compressive stress. There is also a reduction in the mid-sagittal diameter of the vertebral canal. The normal

Study tasks

- Consider the function of the posterior longitudinal ligament in view of the fact that the majority of disc prolapses occur in a posterolateral, rather than a strictly posterior, direction.
- Examine a 'vertebral motion segment' on a flexible plastic spine. If nuclear material escapes into the intervertebral foramen, consider which structure is likely to be affected.
- It has been shown that the hydrophilic nuclear material will swell 2–3 times in size within a few hours if it escapes from the annulus (Dolan *et al.*, 1987). Consider the implications that this has for the patient with a posterolateral disc prolapse.

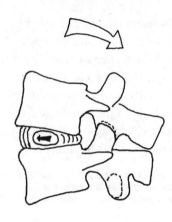

Movement produced by:

- erector spinae
- transversospinales
- interspinales
- intertransversarii
- quadratus lumborum
- assisted by gravity

Figure 7.11 Extension of the lumbar spine (detail of motion segment)

lumbar lordosis is about 54–60° in the standing position, and there is little extra active movement available before the limit of extension is reached. However, the degree of lordosis varies greatly between individuals, and may appear to commence higher or lower than L1. This has considerable functional significance with particular reference to sidebending and rotation movements and the role of the facet joints in these movements.

Further factors affecting the intervertebral discs

Creep

When a load is suddenly applied to a disc but then removed immediately, the disc deforms but recovers relatively quickly with an elastic recoil. However, if the load is maintained, the disc shows a slowly progressive deformation, termed 'creep'. The disc loses height due to fluid expulsion (perhaps 10% over a 4 hour period), which causes annular bulging and this affects the mechanical properties of the motion segment (Adams and Hutton, 1983). Experiments have shown that after a period of loading a motion segment's resistance to flexion falls by about 70%, perhaps due to slackening in the structure of the posterior annulus, and that posterior prolapse is more difficult to simulate (Adams et al., 1987).

(a) Before creep: early in the day/early in Life

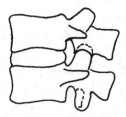

(b) After creep: later in the day/later in Life

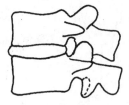

Figure 7.12 The effect of creep on the height of a vertebral motion segment (lateral view) (ligaments are omitted)

 Study tasks

- Using your knowledge of ligaments and apophyseal joints, consider the differences in performance between a 'pre-creep' motion segment and a 'post-creep' motion segment.
- Consider that the pre-creep situation may represent 'youth' and the post-creep situation 'maturity', and try to decide which structures are at risk at each stage.
- Consider what other factors might influence the predisposition to back injury.

Lumbar posture

With reference to the sagittal plane, the presence of a lumbar lordosis is desirable due to its shock-absorbent properties. In the erect position it also probably reduces the hydrostatic pressure in the nucleus. Movement into flexion (particularly unloaded) probably aids the transport of nutrients. An exaggerated lordosis tends to overload the posterior structures, so it seems that neither too much nor too little flexion or extension is desirable, with plenty of movement between segments encouraging optimal nutrition. With these factors in mind, a safe zone has been suggested between approximately 7° of flexion and a little more than 1° of extension, where bending stresses in the spine remain low. Within this zone it is argued that the lumbar motion segment is relatively safe from injury, though any undue loading leads to risk of compressive injury to the vertebral bodies. Active muscular support is optimal in this position, since tendons are particularly effective in absorbing strain energy (Adams, 1994).

Sidebending (lateral flexion)

As in other areas of the spine, it is assumed that sidebending is accompanied by rotation. However, Pearcy and Tibrewal (1984) reported that 'in vitro studies are inconclusive as to the nature of the relation between the two . . . Measurements of lumbar movements in vivo have also demonstrated axial rotation accompanying lateral bending, but . . . the relation between the two has been defined only qualitatively.' In order to obtain more reliable data the authors assessed lumbar movements in two groups of male volunteers using three-dimensional radiography. They found that 'Lateral bending of approximately 10° occurred at the upper three levels,

Study tasks

- Consider the advantages of performing a standing task (e.g. ironing, preparing vegetables, workshop activities) with one foot resting on a low stool, and how this might affect the lumbar spine.
- Consider the benefits of improving the mobility of vertebral motion segments by manipulation.

Study tasks

- Highlight the details shown in Figure 7.13.
- Muscle check.
- Consider the tissues that limit sidebending in the lumbar spine.

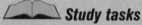 **Study tasks**

- Examine the theoretical possibilities of sidebending with rotation as discussed, using a flexible plastic spine.
- Observe and palpate these movements with a colleague, but do not become discouraged by the difficulties of the task. In assessing your findings, consider the following points (and perhaps some of your own):
- The vertebral column is variable in structure both within and between individuals. For example, anomalies may be present such as asymmetry of apophyseal joint surfaces.
- The presence of the primary and secondary curves (thoracic kyphosis and lumbar lordosis) complicates the relative influence of anterior and posterior parts of the motion segment.
- Lateral curves (scolioses) may be present, which will also complicate movement.
- Muscle and ligament attachments and variations in the performance of these tissues will influence movements and vary both within and between individuals, affected by factors such as health and age.

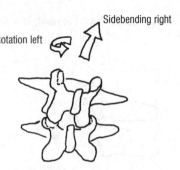

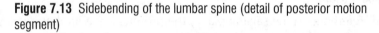

Movement produced by:

- muscles that produce extension of the lumbar spine (Figure 7.11) acting on the same (ipsilateral) side as the movement
- assisted by gravity

Figure 7.13 Sidebending of the lumbar spine (detail of posterior motion segment)

while there was significantly less movement of 6° and 3° at L4–5 and L5–S1 respectively. In the upper lumbar spine axial rotation to the right was accompanied by lateral bending to the left and vice versa. At L5–S1, axial rotation and lateral bending generally accompanied each other in the same direction, while L4–5 was a transitional level.' They also found that in the upper levels sidebending was always accompanied by extension, whereas L4–5 'generally extended but occasionally flexed' and L5–S1 'generally flexed'. For those who wish to pursue the theoretical arguments further, Kapandji's model which emphasises the influence of the anterior motion segment has already been mentioned in Chapter 6 (p. 60), where the work of Lovett and Fryette was also discussed. Fryette (1954) differentiated between thoracic and lumbar movements, but held that the same basic principles apply. These are that if the anterior part of the motion segment (intervertebral discs and bodies) exerts a greater degree of influence on sidebending than the posterior part (the apophyseal facet joints), then accompanying rotation will be towards the convexity of the curve. If, on the other hand, the posterior part of the motion segment has a greater degree of control over the movement, then rotation is guided towards the concavity of the sidebend. The reason for the different outcomes would seem to be whether sidebending commences in a position of flexion, extension or neutral; but it should always be remembered that *in vivo* it is practically impossible to eliminate the effects of the lumbar lordosis and muscular control – factors of which Piercy and Tibrewal (1984) were aware.

It may be noticed that sidebending appears to commence in the lower thoracic spine, and is greater in the upper/mid lumbar, and less in the lower lumbar areas. This is in line with the findings of Piercy and Tibrewal (1984), though it is not easy from palpation alone to be sure about the nature of rotations accompanying the movement.

Rotation

Movement produced by:

- rotatores
- multifidus
- semispinalis thoracis
- obliquus externus abdominis (on the side opposite movement (contralateral))
- obliquus internus abdominis (on the same side as the movement (ipsilateral))

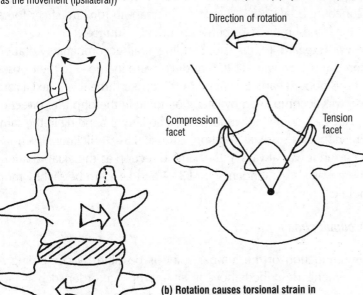

(a) Excessive rotation is restricted by the orientation of the apophyseal joint surfaces (facets) (superior view)

Direction of rotation

Compression facet

Tension facet

(b) Rotation causes torsional strain in the intervertebral disc (anterior view of motion segment)

Figure 7.14 Rotation of the lumbar spine

Study tasks

- Highlight the details shown in Figure 7.14.
- Muscle check.
- Consider the factors that limit rotation in the lumbar spine.

Rotation in the lumbar spine, as elsewhere, appears to be a compound movement involving a degree of sidebending. In the lumbar spine, rotation is limited by the sagittal orientation of the articular processes. The centre of rotation probably shifts, especially if rotation is combined with flexion or extension, but analysis of the movement using an isolated motion segment suggests a centre of rotation that lies in the posterior annulus (Cossette *et al.*, 1971). Rotation about this axis is met by resistance in the apophyseal joints, which form a compression facet and tension facet, accord-

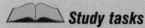

Study tasks

- Observe and palpate rotation in the lumbar spine in a number of seated colleagues. (The sitting position reduces pelvic rotation.)
- Consider whether your findings match the discussion in the text.

Clinical Note

The structural and functional complexity of all vertebral areas means that there is no substitute for careful clinical evaluation of each patient individually, based on the combined skills of observation and palpation, followed by a thorough examination of active and passive movements. This is a skilled and time-consuming process, and perhaps one of the reasons why properly trained manual therapists such as osteopaths are becoming a respected part of modern health care (see for example the proficiency standards published by the General Osteopathic Council, 1999).

ing to the direction of rotation (Adams and Hutton, 1981) (Figure 7.14(a)).

Experiments have shown that only about 1–2° of movement is permitted on each side of a normal segment before damage occurs to the compression facet; but with increasing degeneration greater movement may be observed due to loss of stability (Adams and Hutton, 1981). The importance of this mechanism is that it protects the discs from torsional strain. During rotation, one half of the oblique fibres of the annulus are stretched while the other half are relaxed, according to their orientation. This raises the internal pressure and compresses the nucleus pulposus. There might appear to be a danger of combining flexion with rotation and thereby losing the protective effect of the apophyseal joints, but experiments have suggested that flexion actually decreases the range of axial rotation (Gunzburg *et al.* 1991) due to an increase in forward shear forces which close up the gaps between articular surfaces.

With regard to the sidebending that accompanies rotation, Piercy and Tibrewal's (1984) findings were in line with their observations quoted earlier. That is: 'At the upper three levels, axial rotation was accompanied by lateral bending in the opposite direction … At L4–5, some individuals exhibited lateral bending in the same direction as the axial rotation, and at L5–S1 if lateral bending occurred it was always in the same direction as the axial rotation.' There was also 'a tendency for L3–4 and L4–5 to be slightly more mobile'.

Circumduction

Circumduction of the lumbar spine is possible by combining all movements described above in an appropriate order such as flexion/rotation/sidebending/extension and so on.

The vertebral junction areas

Introduction

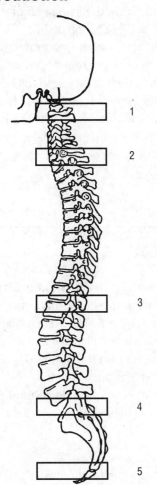

1 Atlanto-occipital junction
2 Cervicothoracic junction
3 Thoracolumbar junction
4 Lumbosacral junction
5 Sacrococcygeal junction

Figure 8.1 The vertebral junction areas

The junction areas of the vertebral column are found at the following points: where the vertebral column commences below the cranium (the craniovertebral junction); where the cervical spine meets the thoracic spine (the cervicothoracic junction); where the thoracic spine meets the lumbar spine (the thoracolumbar junction); and where the lumbar spine meets the sacrum (the lumbosacral junction). In addition, the clinically important sacrococcygeal junction should not be overlooked.

Having seen that the three main areas of the vertebral column are structurally very different, and therefore differ in function as well, it is perhaps not surprising that the junctional areas are of considerable clinical importance; for it is here that a change in both structure and function is to be found. This may be relatively sudden, or more gradual. For example, there is a sudden change in structure between the occiput and C1, and between L5 and the sacrum; but a more gradual change at C7–T1, and at T12–L1.

The clinical significance of these areas is even greater if we also consider the use to which the body may be put, either deliberately or unwittingly. Short-term acute events such as whiplash injuries to the neck, or long-term chronic conditions such as postural hyperextension of the lumbar spine may result in focal injury to the junction areas precisely because the changes in functional demand cannot be accommodated by the structures concerned.

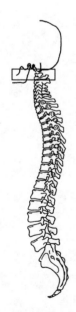

Figure 8.2 The atlanto-occipital junction

The atlanto-occipital junction

The uppermost joint in the vertebral column is the articulation between the occipital condyles and the atlas. The details of this joint have already been outlined in Chapter 5. The vulnerability of this junction lies in the need to combine mobility (which the head requires in order to orientate the senses) with stability.

The average adult human head weighs approximately 5–7 kg, and the constant need for orientation atop a flexible column presents a real structural challenge; this is achieved by means of the following:

- the shape of the bone structures
- the position and number of the supporting ligaments
- muscular support
- the adaptation provided by the postural curves

Clinical Notes

Vulnerability of the junction

The sheer number of joints and ligaments present in this junctional area (and the adjacent atlanto-axial joints should not be overlooked) means that this region is often susceptible to systemic disorders such as rheumatoid arthritis. It is beyond the scope of this book to explore rheumatological disorders, but in terms of traumatology a common injury seen in the upper cervical spine is the acceleration trauma known as whiplash injury. This may present as a flexion/extension injury to a passenger in a stationary vehicle which is struck from behind. At the moment of impact the passenger's body is flung forward, but the inertia of the head means that it remains behind, and is forced into rapid hyperextension, which will tend to damage the anterior structures of the neck and cause damage to the disc and apophyseal joints. Sudden and total extension may also fracture the posterior arch of the atlas. The head may then be flung forwards into flexion by the continuing momentum. Of course, much depends on the force of impact and position of the head at the time. The role of a properly placed headrest is of crucial importance in preventing injuries of this nature.

Compression traumas such as diving accidents or severe impact in contact sports may produce similar effects to whiplash injuries if forced extension occurs.

Conversely, forced flexion trauma may occur at the craniovertebral junction, resulting in injury to the posterior structures, possible apophyseal joint dislocation and ultimately fracture of the vertebrae with consequent neurological damage.

In contrast, a case of death by hanging involves a fatal traction effect induced by unrestrained body weight falling away from a restrained head. The weight of the body forces the dens to rupture the transverse ligament and then the spinal cord, as the unsupported body falls away.   *(cont. p. 82)*

 Study tasks

- Review the anatomy of the atlanto-occipital joint in Chapter 5 with particular reference to the bones, ligaments, and muscles.
- Consider how these structures individually meet the needs of support and mobility.

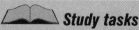

Study tasks

- Using your knowledge of movements in the upper cervical spine consider again the compensatory adjustments that are likely at C1–C2 and C0–C1 if a scoliosis develops in the vertebral column.
- With reference to Chapter 5, consider which muscles are likely to become hypertonic. Also review the course of the vertebral artery and the greater and lesser occipital nerves and consider the possible significance of the proximity of muscles such as the rectus capitis posterior minor to the dura mater (the outer covering of the brain and spinal cord). Consider the relevance of your findings to the condition popularly known as 'tension headache'.

(cont.)

In all these cases it will be apparent that the severity of injury depends on a number of variables, which include the direction and degree of force, the position of the body, and so on. An underlying knowledge of the detailed anatomy of this area is important in making an accurate diagnosis of the damaged tissues and assessing the various treatment options, as appropriate.

In more chronic postural disorders, the upper cervical joints are also a site of 'final adjustment' for discrepancies such as scolioses. The eyes must always be level with the horizon for proper orientation of the senses. The special ability of the upper cervical joint surfaces to adjust and compensate for postural discrepancies may produce strain patterns in muscles and ligaments with consequent symptoms of strain. The neurovascular supply to the head (for example, the vertebral artery, greater and lesser occipital nerves, and proximity of the dura mater) may be compromised by the chronic hypertonicity of certain muscle groups such as the suboccipital and cervical extensor muscles, the trapezius and sternocleidomastoid.

The cervicothoracic junction

The presence of the first rib and its attachments to the first thoracic vertebra (T1) gives obvious structural definition to the cervicothoracic junction at C7–T1. However, their prominent spinous processes mean that C7 (vertebra prominens) and T1 are often confused on palpation. Other differences are more obvious when bone specimens are examined. Thus, T1 lacks uncinate processes and a foramen transversarium, and it possesses larger club-shaped transverse processes than do the cervical vertebrae. Also, at T1 the inferior vertebral articular processes begin to slope downwards to become more coronal in orientation, which allows the upper thoracic spine to rotate freely, but see Clinical Note opposite.

The immense influence of the thoracic (rib) cage means that there is a considerable change in function imminent at the cervicothoracic junction, heralded by a change in direction of the anteroposterior spinal curves from cervical lordosis to thoracic kyphosis.

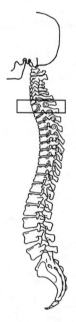

Figure 8.3 The cervicothoracic junction

Clinical Note

Vulnerability of the junction

The greater mobility and therefore vulnerability of the upper cervical spine mean that acute trauma is generally more damaging to the cervical spine than to the cervicothoracic junction itself, apart from accidents involving certain inverted gymnastic manoeuvres. The commonest dysfunction in this area is probably the result of chronic postural habits that encourage the so-called 'rounded shoulders' effect. Rather than being caused by the shoulders alone, this is usually the result of an exaggerated thoracic kyphosis which has the effect of carrying the shoulders anteromedially. The effects of this exaggeration in curvature is to force the apophyseal joint surfaces (the articular processes) to lie in a more horizontal plane, which restricts their normal range of movement and causes them to become excessively weightbearing (see Figure 6.11). This may accelerate degenerative changes, with associated pain and discomfort in the lower cervical spine, where a change in direction of posture from flexion (thoracic) to extension (cervical) now becomes sudden rather than gradual. Another unfortunate consequence of 'rounded shoulders' is that the position of the scapula becomes protracted on the rib cage, which distorts the supportive muscles such as the levator scapulae, the rhomboids and upper trapezius. This may lead to neck and shoulder pain which has been described as 'scapulocostal syndrome' (Cailliet, 1991a). Scapular displacement also results in an altered glenohumeral position which may have unfortunate consequences for glenohumeral movement, especially abduction. This point is explored further in Chapter 10.

 Study tasks

- Obtain and compare bone specimens of C7 and T1.
- With reference to Chapters 5 and 6, review the differences between cervical and thoracic movements.

 Study tasks

- Obtain a flexible plastic spine and explore the effects of an increase and decrease in anteroposterior curvature on movements at the cervicothoracic junction.
- Consider all possible physical and psychological factors that may result in 'rounded shoulders' and/or exaggerated thoracic kyphosis.

The thoracolumbar junction

The traditional view of the junction between the thoracic and lumbar spine has sometimes been that there is a sudden change in

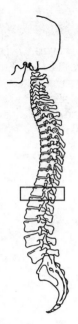

Figure 8.4 The thoracolumbar junction

the orientation of the articular processess between T12 and L1. This view suggests that the superior articular processes of T12 lie in a typically thoracic coronal plane, while the inferior processes that articulate with the superior processes of L1 show a typically lumbar, sagittal, orientation.

However, some doubt has been cast on this assertion. A comprehensive study of computerised tomography archives has suggested a gradual or progressive change in the orientation of the articular processes from T10–11 to T12–L1 in the majority of cases studied. Only 30% showed the sudden change of orientation at T12. Interestingly, the more gradual transition also seemed twice as likely to occur in females (Singer *et al.*, 1989).

Clinical Notes

Vulnerability of the junction

It has already been suggested (Chapters 6 and 7) that the coronal orientation of the thoracic apophyseal joint surfaces encourages rotation, whereas the more sagittal orientation of the lumbar apophyseal joints facilitates flexion/extension. If an abrupt change were to occur at T12, the superior articular surfaces would tend to encourage rotation, whereas the inferior articular surfaces would not. This could lead to a danger of strain on rotation movements. In fact Singer (1994) suggests that gradual transition is likely to be more effective in limiting rotational strain at the junction of the rotationally more mobile thoracic region as it meets the rotationally less mobile lumbar region. This may be especially relevant in an activity such as walking, when in order to keep the eyes forward, the thorax rotates in the opposite direction to the lumbar spine and pelvis. Singer also suggests that this may explain the preponderance of an apparently gradual transition in females, where a wider pelvis coupled with deeper lumbar lordosis would make a sudden transition mechanically more stressful. Thus, a gradual transition would act as a damping mechanism, or brake, to vertebral rotation. Singer (1994) also mentions evidence of enlarged lumbar mamillary processes around the thoracolumbar junction, and suggests that rotatory muscles such as the multifidus which attach to the mamillary processes could in fact act to counteract sudden rotation strains. Thus, the real significance of this vertebral junction may well lie in its dynamic role as a buttress against rotational strain as thoracic meets lumbar; and the sorts of injuries seen at this level may reflect a breakdown in coping mechanisms, in either an acute or a chronic context. As Singer (1994) advises, in view of these findings, caution should be exercised by manipulative therapists in the application of rotational techniques to this area.

 Study task

- Try to obtain bone specimens to study the change in orientation of articular surfaces at the thoracolumbar junction.

The lumbosacral junction

The junction between the lowest lumbar vertebra (L5) and the upper surface of the sacrum (denoted as S1) is structurally distinct, and this is reflected in function and use. This is where the lumbar lordosis, or secondary lumbar extension curve, meets the fixed primary curve of the sacrum. The result is a 'lumbosacral angle' of about 30–50° as defined in Chapter 7 (see Figure 7.6).

Movements at the lumbosacral junction

Čihák (1970) has drawn attention to the variations that occur in the orientation of the articular apophyseal joint surfaces at the lumbosacral junction. Their tendency to lie in the coronal plane discourages shear, but according to Čihák variations are common, even between joints in the same individual. Flexion/extension is the 'purest' movement at the lumbosacral junction, and in childhood up to three-quarters of this lumbar movement may occur here, but gradually L4–5 becomes proportionately more significant (Williams, 1995). The combined rotation/sidebending movements found in the lumbar spine, including the lumbosacral junction, are discussed in Chapter 7. It is perhaps worth restating that the work of Piercy and Tibrewal (1984) suggests that, whereas rotation and sidebending occur to opposite sides in the upper lumbar spine, these movements tend to accompany each other in the same direction at the lumbosacral junction, with L4–5 being transitional.

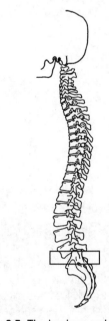

Figure 8.5 The lumbosacral junction

Clinical Note

Vulnerability of the junction

A most important feature of the lumbosacral junction is that the base of the sacrum forms the foundation on which the vertebral column rests. Cailliet (1981) likened L5 to a box resting on a slope, and gravity naturally produces a tendency for the vertebral column above, to slide forwards. At the lumbosacral junction the angle results in a shear force that is opposed by the bony structure of the vertebrae, the ligaments, the intervertebral discs and the supporting muscles. However, as the angle increases so does the shear force, and in mechanical terms a 50° angle causes a shear force equivalent to 75% of the weight of the moveable object (Cailliet, 1981).

Study task

- Consider which features of the vertebrae and which ligaments effectively oppose shear at the lumbosacral junction.

Study tasks

- Revise the structures of the lumbar spine in Chapter 7, and then consider why a person with an exaggerated lumbar lordosis is likely to experience low back pain: (i) sitting slumped in a soft chair; (ii) painting a ceiling.
- Consider the implications of this for: (i) seating design; (ii) advice in the workplace.
- Consider the necessary criteria for designing suitable seating for use in schools, offices, factories, homes and vehicles.

Within normal limits (approximately 30–50°) the lumbosacral angle reflects the wedged shape of the body of L5 and the wedged shape of the lumbar intervertebral discs. However, an undesirable increase in the lumbosacral angle will increase the shear force and will also increase, or perhaps be the result of, the lumbar lordosis, which in turn increases the shear strain on the upper lumbar vertebrae and associated structures. The pelvis may anteriorly rotate, and this may, in turn, produce compensatory flexion at the hips and knees.

Long-term postural changes such as these may produce dysfunctional effects. For example, there may be shortening in the capsular, ligamentous, fascial and muscular structures, compression and degenerative changes in the posterior aspects of discs, compressive loading of apophyseal joints, approximation of spinous processes (the so-called 'kissing spines' seen in radiographs), and narrowing of intervertebral foramina with consequent nerve root irritation.

The main movements of flexion and extension tend to be focused at the 'hinge points' of L4–5 and the lumbosacral junction, with a tendency to overload these joints in lifting activities (see Chapter 7).

However, the unfortunate tendency in so-called civilised societies to sit in chairs rather than squat (the latter being healthier, if less practical) and to drive vehicles makes low back pain an almost inevitable consequence of prolonged periods of weightbearing imposed on a flexed lumbosacral junction.

The sacrococcygeal junction

The sacrococcygeal junction represents the joint between the apex of the sacrum and the base of the coccyx, which is a secondary cartilaginous joint, or symphysis, but has elliptical articular surfaces permitting a small range of flexion/extension and an almost insignificant amount of lateral movement.

(a) Flexion of the sacrococcygeal joint (lateral view)

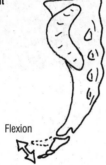

Flexion

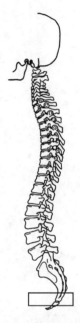

Figure 8.6 The sacrococcygeal junction

(b) Ligaments (anterior view)

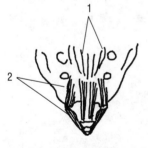

(c) Ligaments (posterior view)

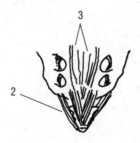

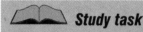

Study task

- Shade the ligaments shown in Figure 8.7(b–c) in colour.

1 Anterior (ventral) sacrococcygeal ligament
2 Lateral sacrococcygeal ligament
3 Posterior (dorsal) sacrococcygeal ligament – deep fibres from posterior longitudinal ligament

Figure 8.7 The sacrococcygeal joint

The coccyx is really the vestigial remnants of a tail. Its ability to flex and extend enhances the influence of the sacrum to increase and decrease the pelvic outlet (see Chapter 9). These accommodatory movements take place during activities such as defecation and parturition, and the attachment of the pelvic diaphragm muscles means that some active movement is possible.

Clinical Note

Pelvic dysfunction

The main vulnerability of the coccyx lies in its superficial, and in some cases virtually subcutaneous, position, which can make either sitting on hard surfaces uncomfortable or a fall in a position of flexion a real hazard. The coccyx may fracture in a fall, or be forced into an unnaturally flexed position. Pain in the coccyx (coccydinia, coccygodinia) should always be evaluated by a qualified practitioner.

Chapter 9

The pelvis and sacro-iliac joints

Introduction

The pelvis is a specialised bony ring consisting of two halves (the innominate bones), united anteriorly by the pubic symphysis and posteriorly by the wedge-shaped sacrum, which has been likened to the keystone in an arch.

The pelvic ring may be divided into a posterior weightbearing arch running from the sacrum to the acetabular sockets of the hip, and an anterior pubic arch which acts as a 'tie beam' to resist tension or compression depending on whether forces are pulling apart or pushing together.

Each innominate consists of three separate bones (the ilium, pubis and ischium), which are joined by cartilage in the infant but ossify in the adult. They meet in the bony hip socket (acetabulum), hence the important fact that the pelvis absorbs the locomotive forces from the lower limbs below and the weight of the body above. Since movement is essential, the pelvic ring must be able to accommodate these forces. At the same time it must support and protect the abdominal viscera and in women of reproductive age it has important obstetric functions.

The bones

(a) Anterior view

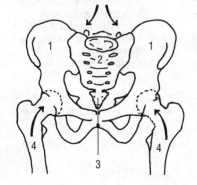

Weightbearing forces

Bones
1 Innominate
2 Sacrum
3 Coccyx
4 Femur

(b) Posterior view

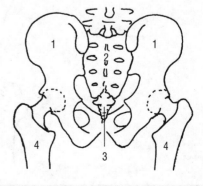

Study tasks

- Lightly shade the three separate parts of the innominate bone in different colours (Figure 9.1(c)) and note the names of the bony features shown.
- With reference to the list of functions of the pelvis, match each with one or more of the features shown in the diagrams:
  (i) protection and support of abdominal organs (ii) distribution of forces from above and below (iii) links vertebral column and lower limbs (iv) muscle attachment (v) obstetric functions in females.

(c) The right innominate bone (lateral aspect)

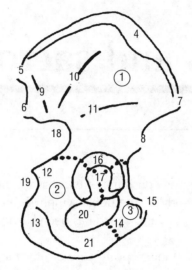

① Ilium
② Ischium
③ Pubis

 Limits of iliac, pubic and ischial bones

4 Iliac crest
5 Posterior superior iliac spine
6 Posterior inferior iliac spine
7 Anterior superior iliac spine
8 Anterior inferior iliac spine
9 Posterior gluteal line
10 Anterior gluteal line
11 Inferior gluteal line
12 Ischial spine
13 Ischial tuberosity
14 Pubic ramus
15 Pubic tubercle
16 Acetabulum
17 Acetabular notch
18 Greater sciatic notch
19 Lesser sciatic notch
20 Obturator foramen
21 Ischial ramus

Figure 9.1 The bones of the pelvis

The joints and ligaments

The main ligaments are shown diagrammatically, but their functional significance will be discussed in the sections on the joints.

(a) Ligaments of the pelvis (anterior view)

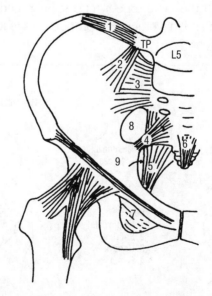

1 Ilio-lumbar
2 Lumbosacral
3 Ventral sacro-iliac
4 Sacrospinous
5 Sacrotuberous
6 Ventral sacrococcygeal
7 Obturator membrane
8 Greater sciatic foramen
9 Lesser sciatic foramen
TP Transverse process of L5

(b) Ligaments of the pelvis (posterior view)

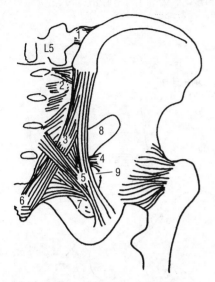

1 Ilio-lumbar
2 Short dorsal
   sacro-iliac
3 Long dorsal
   sacro-iliac
4 Sacrospinous
5 Sacrotuberous
6 Dorsal sacrococcygeal
7 Obturator membrane
8 Greater sciatic foramen
9 Lesser sciatic foramen

Study tasks

• Lightly shade the ligaments in colour and highlight the details shown.

Figure 9.2 Ligaments of the pelvis

The pelvic joints (general)

The preceding discussion implies that the pelvis accommodates a number of diverse activities, which could vary from protecting an unborn foetus to winning an Olympic medal for hurdling (possibly in the same individual!). In order to combine the need for strength and stability with that for mobility, the precise function of these joints is complex and defies easy analysis. Perhaps the most striking characteristic is that of a strong mechanism of interlocking joint surfaces, which by definition allows some movement but which is stabilised by powerful ligaments.

The highest point of the functional posterior arch is the base of the sacrum and the lumbosacral joint. At this point the weight of the body from above tends to force the sacrum downwards between the innominate bones, and forwards due to the lumbar lordosis and lumbosacral angle (see Chapter 7). These movements are resisted by a number of factors:

• the wedged shape of the sacrum;
• the shape and interlocking nature of the auricular (ear-shaped) sacro-iliac joint surfaces;
• the anterior arch and pubic symphysis;
• the strong ligaments (including the powerful interosseous ligament that underlies the dorsal sacro-iliac ligaments);

- the surrounding muscles and fascia, especially the gluteus maximus, piriformis and biceps femoris, which are partially attached to the sacrotuberous ligament (see Figure 9.2).

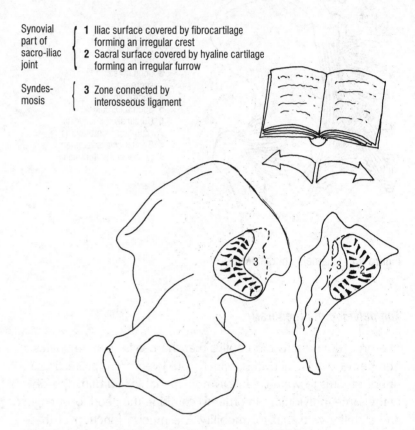

Synovial part of sacro-iliac joint
{
  1 Iliac surface covered by fibrocartilage forming an irregular crest
  2 Sacral surface covered by hyaline cartilage forming an irregular furrow
}

Syndes-mosis
{
  3 Zone connected by interosseous ligament
}

Figure 9.3 The sacro-iliac joint – opened out (like a book)

Figure 9.3 presents a simplified view of the sacro-iliac joint, which is regarded as part synovial joint, part syndesmosis. Kapandji (1974) and Williams (1995) divide the synovial part of the sacro-iliac joint into three distict sections. The articular surfaces are somewhat sinuous in the anterior, or upper segment at S1 level, and display a sacral concavity, or furrow, fitting into an iliac convexity at the middle level (S2), which is less noticeable in the lower or posterior segment. The function of this interlocking mechanism appears to be to resist the downward and forward components of shear which were discussed with reference to the lumbosacral angle in Chapter 7. The interlocking middle segment probably plays a key role in this; with the complimentary shapes of the upper and lower segments allowing a rotatory movement, which will be discussed in greater detail below. The apparently complex structure of the joint, which includes interlocking fibrous components with more mobile synovial features, cleverly accommodates both weightbearing needs and stability with the requirements of movement.

Movements at the sacro-iliac joints

There have been many attempts to describe and analyse the complexities of movement at the sacro-iliac joint. It is difficult to obtain precise measurements, and there are obvious variations between the sexes, as well as variations between age groups. Movement becomes more restricted with age (Williams, 1995), but in females ligamentous laxity under hormonal influence results in palpably mobile sacro-iliac joints. This is not to deny that movement can also be palpated in males. There have been a number of different theories about the nature of sacro-iliac movement, of which only a limited account will be given here. The interested reader is referred to the extensive literature on the subject, some of which is discussed in Kapandji (1974) and Lee (1989).

Kapandji (1974) classifies the various theories of sacral movement under the headings 'nutation' and 'counternutation' which are terms derived from the Latin verb *nutare* ('to nod'). This is a useful classification, since a general forward and backward movement of the sacrum does seem to occur, even though there may be some disagreement about where the axis of movement lies or whether the movement involves rotation or slide. Weisl (1955), for example, argues in favour of a transverse axis of rotation which lies outside the joint, as much as 10 cm below the sacral promontory. This was seen as a mobile axis which results in an impure swing between the articular surfaces.

Fryette (1954) suggested three different types of sacrum in terms of their anterior-posterior dimensions, variations in shape of the auricular surface and variations in orientation of the S1 articular processes. Factors such as these make generalisation difficult.

As in certain movements of the lumbar spine (see Chapter 7), this may be another topic where both student and practitioner are better served by individual observation and palpation than by being drawn into generalisation. Also, it seems unfortunate that many of the theories of sacroiliac movement make little or no mention of the influence of muscle attachments, which would seem to be highly influential in exerting differential forces on the pelvic bones, either directly or via their ligaments. This is another characteristic that should encourage individual clinical examination.

Nutation

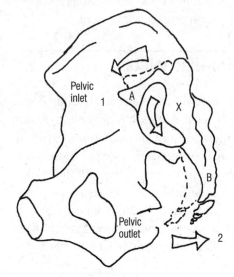

1 Promontory of sacrum (A) moves anteriorly and inferiorly
2 Apex of sacrum (B) and coccyx move posteriorly
X Axis of rotation (approximately S2)

Figure 9.4 The movement of nutation at the sacro-iliac joint (sagittal view)

This movement, which is essentially passive, implies that the sacral promontory moves forwards and downwards, while the apex (tip) of the sacrum as well as the coccyx move backwards. The sacral promontory may move 5–6 mm (Williams, 1995). Since the sacrum moves between the iliac bones, these are brought closer together while the ischial tuberosities move further apart. The pelvic outlet increases in size, while the pelvic inlet decreases, and this corresponds to the expulsive phase during parturition (Kapandji, 1974). This is also the forward position that the sacrum adopts during standing, and is limited by the interlocking nature and shape of the auricular surfaces plus tension in the anterior sacro-iliac ligaments and sacrotuberous and sacrospinous ligaments.

The role of muscles is uncertain, but the coccygeus muscle and the sacrospinous ligament appear to merge, and it is likely that muscles such as the gluteus maximus, piriformis and biceps femoris, which have partial attachment to the sacrotuberous ligament, provide dynamic support to the pelvis.

Counternutation

This is the return movement, or 'backward nodding' of the sacrum, in which the promontory of the sacrum moves upwards and back-

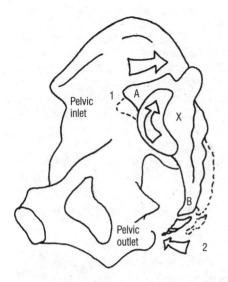

1 Promontory of sacrum (A) moves superiorly and posteriorly
2 Apex of sacrum (B) and coccyx move anteriorly
X Axis of rotation (approximately S2)

Figure 9.5 The movement of counternutation at the sacro-iliac joint (sagittal view)

wards limited by tension in the posterior and anterior ligaments. The iliac bones move apart, the ischial tuberosities move closer together and the pelvic outlet diminishes in size but the inlet increases. This is said to correspond to the early stages of parturition (Kapandji, 1974).

 Study tasks

• Shade the position of the sacrum in counternutation.
• Palpate the small movements of nutation and counternutation by placing a colleague prone on a comfortable surface (pillow under abdomen) and placing the palmar surface of your palpating fingers over the upper sacrum. Exert only light pressure with your other hand over the spinous tubercles (processes) of the lower sacrum. Only a very light pressure is needed to 'spring' the sacrum.
• Repeat the movements, but with your palpating hand on each sacro-iliac joint in turn (just medial to the easily palpable posterior superior iliac spines). Try to differentiate between the degree of mobility in each joint. This may seem difficult at first, but your palpatory skills will improve with practice. Try to locate the sacrotuberous ligament by palpating through the relaxed gluteus maximus muscle. The ligament is palpable as it passes obliquely downwards and outwards from the sacrum to the ischial tuberosity.

Clinical Notes

Pelvic dysfunction

The assumption that the sacrum moves into nutation and counternutation also implies that the innominate bones on either side move in the opposite direction relative to the sacrum. There is also no reason to suppose that they cannot move independently on the sacrum. This possibility has given rise to some interesting and complex descriptions of various sacro-iliac dysfunctions termed 'lesions' by classical osteopaths such as Downing (1923), who described the following possible scenarios (quoted in Sammut and Searle-Barnes (1998): (cont. p. 96)

(cont.)

- unilateral anterior innominate (unilateral sacral counternutation);
- unilateral posterior innominate (unilateral sacral nutation);
- bilateral posterior innominates (bilateral sacral nutation);
- bilateral anterior innominates (bilateral sacral counternutation);
- unilateral anterior innominate (unilateral sacral counternutation);
- true twisted pelvis: combined unilateral posterior/unilateral anterior innominate on the opposite side;
- pseudo-twisted pelvis: apparent asymmetry of the pelvis due to imbalanced lumbar muscle tension and L5 rotation; there will be a resultant compensatory functional scoliosis probably due to the tension pull of the iliolumbar ligament and the contracted state of the surrounding musculature.

Downing then proceeded to describe various manipulative procedures designed to correct these lesions or 'subluxations'. Readers who are interested in exploring and evaluating these techniques are advised to consult the original texts, for example Downing (1923), or more recent contributions such as Stoddard (1980; 1983), Bourdillon et al. (1992) and Hartman (1997), among others.

The role of muscles in pelvic movements

There is a tendency to assume that only passive movements are possible at the sacro-iliac joint because the surrounding muscles do not actively move the joints. This may be misleading, since it tends to underestimate the role of muscle activity in the region generally. Thus, active movements of the lumbar spine and hip produce predictable movements in the pelvis. For example, in active lumbar flexion from the standing position, there is an initial counternutation of the sacrum as the innominates move anteriorly, although eventually the entire pelvis, including the sacrum, is rotated anteriorly in the fully flexed position. Conversely, in lumbar extension there is a tendency towards nutation of the sacrum as the innominates rotate posteriorly. The innominates are greatly influenced by hip movements, and may also be under more direct muscular control.

This may be seen for example during the 'gait cycle' (see p. 188), which is conveniently divided into 'heel strike', 'midstance' and 'push off/toe off'. At heel strike, the gluteus maximus and the hamstrings tend to pull the innominate posteriorly. At midstance, the abductor muscles stabilise the innominate on the weightbearing side, but as 'toe off' is reached, tension in the iliopsoas results in anterior rotation of the innominate. Thus, a number of powerful muscles are attached to the innominate bones, and these may exert a powerful effect pulling the innominates either anteriorly or posteriorly. The sacrum also has significant muscle attachments such as to the piriformis and gluteus maximus. Whether the sacrum or innominates remain 'stuck' at the end of these types of movement may depend on the amount of force or trauma applied.

Clinical Notes

Dysfunction in the sacro-iliac joints may be acute, and the result of a sudden flexion or rotation movement or fall, which can tear posterior ligaments and muscles and in extremis can dislocate the joint surfaces. In this context it is also prudent to consider the possibility of injury to the pubic symphysis. Stoddard (1983) describes a case of a 3 cm separation at the pubic symphysis of a man who fell from a horse, even though his sacro-iliac joints were apparently asymptomatic. On the other hand there may be chronic dysfunction (perhaps following an acute injury) involving either posterior or anterior rotation of either innominates or sacrum, as described by Downing (1923). Problems often arise as a result of one or more of the following reasons:

- the permitted range of movement at the joint surfaces is only small;
- the surrounding muscles are extremely powerful;
- laxity of pelvic ligaments may encourage hypermobility in the joints;
- in childbearing women, hormone release at parturition actively encourages hypermobility.

Any one, or several of these factors may result in either one or both of the sacro-iliac joint surfaces becoming jammed or restricted in an end-of-range position, presumably due to irregularities in the joint surfaces. The onset may be sudden, perhaps due to a

Study tasks

- Obtain a plastic model of the pelvic bones, or refer to an articulated skeleton, and consider the effects on apparent leg length of a posterior rotation and an anterior rotation of either the right or left innominate bone of the pelvis.
- Palpate the posterior superior iliac spines of a standing colleague (the 'dimples' either side of the sacrum). If you carefully position your thumbs they should appear level. If they are not, consider the possibility of an anterior or posterior rotation of the innominate bones.

fall, rapid rotation of the pelvis especially under load, or in childbirth. Or the onset may be gradual, perhaps due to increased tone of one of the powerful muscle groups, such as the hamstrings, which attach to the ischial tuberosities of the innominate bones. Adjustive manipulation has often had successful results in the treatment of sacro-iliac joint problems, especially if the joint is 'stuck'. However, the role of hypertonic muscle activity in gradually shifting the sacrum or innominate out of position should not be overlooked despite the more spectacular adjustive treatments.

Chapter 10

The shoulder

Introduction

The shoulder is an articular girdle of bones, muscles and joints, whose main function is to allow the hand maximum spatial versatility. In order to achieve this, the range of the spheroidal glenohumeral joint is increased by the movements of the scapula, by the swivel action of the strut-like clavicle and also by a degree of flexibility available in the thoracic spine. The neurovascular supply to the shoulder is also derived from the cervical and thoracic regions.

It is important, therefore, that all of these elements in the shoulder girdle are able to function efficiently for proper shoulder movement to take place. It also follows that a proper clinical evaluation of the shoulder should pay attention to the mechanics of both the cervical and thoracic spine, as well as to the functional components of the shoulder girdle itself.

The bones

(a) Anterior view

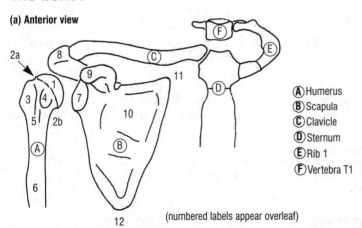

Ⓐ Humerus
Ⓑ Scapula
Ⓒ Clavicle
Ⓓ Sternum
Ⓔ Rib 1
Ⓕ Vertebra T1

(numbered labels appear overleaf)

Study tasks

- Shade the main bones in colour.
- Highlight the names of the bone features.
- Obtain bone specimens if possible, and identify the features shown.
- Identify the main bones on a colleague using palpation. Try to note the features that are palpable, and those that are not.

(b) posterior view

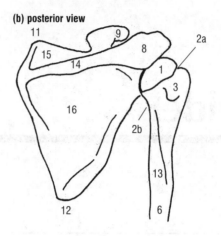

1 Head
2a Anatomical neck
2b Surgical neck
3 Greater tuberosity
4 Lesser tuberosity
5 Intertubercular sulcus (for long head of biceps brachii muscle)
6 Shaft
7 Glenoid cavity
8 Acromion
9 Coracoid process
10 Costal surface
11 Superior angle
12 Inferior angle
13 Sulcus for radial nerve
14 Spine
15 Supraspinous fossa
16 Infraspinous fossa

Figure 10.1 The bone features of the shoulder girdle

The joints and ligaments

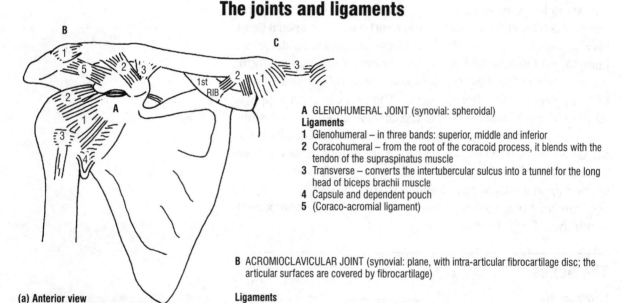

(a) Anterior view

A GLENOHUMERAL JOINT (synovial: spheroidal)
Ligaments
1 Glenohumeral – in three bands: superior, middle and inferior
2 Coracohumeral – from the root of the coracoid process, it blends with the tendon of the supraspinatus muscle
3 Transverse – converts the intertubercular sulcus into a tunnel for the long head of biceps brachii muscle
4 Capsule and dependent pouch
5 (Coraco-acromial ligament)

B ACROMIOCLAVICULAR JOINT (synovial: plane, with intra-articular fibrocartilage disc; the articular surfaces are covered by fibrocartilage)

Ligaments
1 Acromioclavicular
2 Trapezoid } Coracoclavicular – limiting protraction and retraction of the clavicle but aiding
3 Conoid } posterior rotation when scapula forwardly rotates

C STERNOCLAVICULAR JOINT (synovial: modified saddle, with intra-articular fibrocartilage disc; the articular surfaces are covered by fibrocartilage; the clavicle also articulates with the first costal cartilage)
Ligaments
1 Anterior sternoclavicular
2 Anterior and posterior costoclavicular
3 Interclavicular (may include small particles of bone (ossicles))

Figure 10.2(a) Joints and ligaments of the shoulder girdle

Study tasks

- Shade the ligaments in colour.
- Highlight the names and details of the features shown.
- Observe and palpate the joints on a colleague. Note any differences between right and left sides, and try to identify the subject's dominant hand by observation and palpation only.

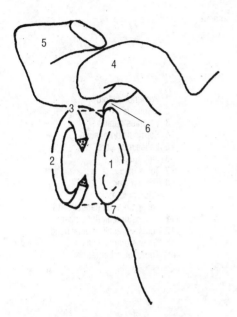

1 Glenoid cavity of the scapula –
  covered by hyaline cartilage
  thicker at the margins
Note: The hyaline articular cartilage
covering the head of the humerus is
reciprocally thinner at its margins
2 Glenoid labrum – a fibrocartilage
  rim which deepens the cavity
3 This part blends with the tendon
  of the long head of the biceps
  brachii muscle, and attaches to
  the supraglenoid tubercle
4 Coracoid process
5 Acromion
6 Supraglenoid tubercle
7 Infraglenoid tubercle

Figure 10.2(b) Oblique view of glenoid surface of scapula to show glenoid labrum (detached and cut)

Muscles and movements

The function of muscles is not only to move but also to protect and support. This is exemplified in the shoulder, and observation will show that the superficial muscles give shape and form to the shoulder, as well as providing outer protection.

(a) The superficial muscles: anterior aspect

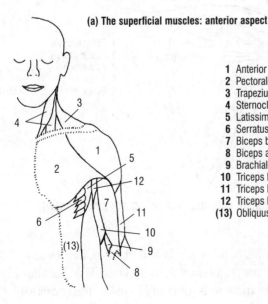

1 Anterior deltoid
2 Pectoralis major
3 Trapezius (upper fibres)
4 Sternocleidomastoid
5 Latissimus dorsi
6 Serratus anterior
7 Biceps brachii
8 Biceps aponeurosis
9 Brachialis
10 Triceps brachii (medial head)
11 Triceps brachii (lateral head)
12 Triceps brachii (long head)
(13) Obliquus externus abdominis

Study task

• Identify on a colleague the
  superficial muscles shown.

(b) The superficial muscles: posterior aspect

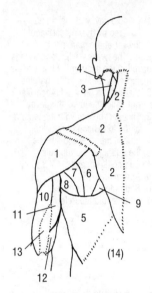

1 Posterior deltoid
2 Trapezius
3 Splenius capitis
4 Sternocleidomastoid
5 Latissimus dorsi
6 Infraspinatus
7 Teres minor
8 Teres major
9 Rhomboid major
10 Triceps brachii (lateral head)
11 Triceps brachii (long head)
12 Triceps brachii (medial head)
13 Triceps brachii tendon
(14) Thoracolumbar fascia

Figure 10.3 The superficial muscles of the shoulder and surrounding muscles

Lying deep beneath the superficial muscles are the musculo-tendinous 'rotator cuff' muscles, which are particularly important in providing support and integrity to the glenohumeral joint.

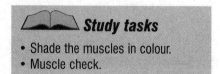

Study tasks

• Shade the muscles in colour.
• Muscle check.

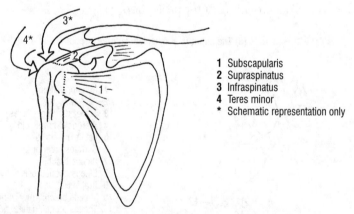

1 Subscapularis
2 Supraspinatus
3 Infraspinatus
4 Teres minor
* Schematic representation only

Figure 10.4 The rotator cuff muscles (schematic anterior view)

The active roles of the scapula and clavicle, as well as the cervical and thoracic areas, mean that there are a large number of muscles that act over the entire functional area of the shoulder.

In order to minimise tendon friction over these surfaces there are a number of bursae present. The most important are:

- the bursa between the tendon of the subscapularis and the shoulder capsule, which emerges through a gap in the capsule between the superior and middle bands of the glenohumeral ligament;
- the subacromial bursa between the deltoid and the capsule, which also separates the tendon of supraspinatus from the arch of the acromion and coraco-acromial ligament (Figure 10.11);
- a bursa overlying the surface of the acromion;
- others may be found near the tendinous insertions of the infraspinatus and latissimus dorsi, between the teres major and the long head of the triceps, and in the vicinity of the coracoid process and its muscle attachments.

It is also important to note in this context that the tendon of the long head of the biceps brachii is ensheathed by synovial membrane in order to reduce friction as it emerges from the capsule of the glenohumeral joint. It then passes through the intertubercular sulcus of the humerus, held in position by the transverse ligament (Figure 10.2 (a)). Note that during movements of the shoulder it is the humerus that moves relative to the tendon, and not vice versa.

Before examining shoulder movements as a whole, it is useful to consider the individual movements of the scapula, humerus and clavicle, and these will be summarised in turn.

Movements of the scapula

Examination of the structure of the scapula reveals the function of this bone, which is commonly called the 'shoulder blade'. It is in fact more like a sculpted shield, with grooves and bony processes that provide generous attachment points for the powerful muscles that support and coordinate the movements of the shoulder girdle. All the muscles that act directly on the shoulder (except one), are attached to the scapula.

The sequence of diagrams in Figure 10.5 shows the movements of the scapula and lists the muscles that produce these movements. Some of these movements do have certain independent value. For example, elevation and depression of the scapula are social signals. But the main purpose of forward and backward rotation and protraction and retraction is to slide the scapula around the rib cage in order to increase the range of movement of the glenohumeral joint.

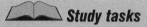 **Study tasks**

- Obtain a bone specimen for reference, and then attempt to palpate the outline and borders of the scapula on a colleague. Identify those features that are palpable and those that are not. Remember to palpate laterally and anteriorly as well.
- Consider which important shoulder muscle is *not* attached to the scapula.
- Observe the independent movements of the scapula on a colleague with their arms by their sides. Try to discover these movements before studying Figure 10.5 (note that one type of scapular movement is difficult without elevating the arm).

(a) Elevation of the scapula

Movement produced by:
• trapezius (upper fibres)
• levator scapulae

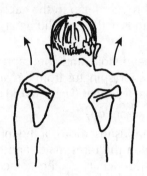

(b) Depression of the scapula

Movement produced by:
• gravity
• serratus anterior (lower fibres)
• pectoralis minor

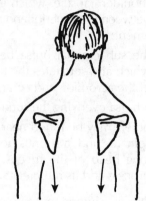

Elevation of the scapula occurs when shrugging the shoulders and depression occurs either with gravity when the shoulders are relaxed, or actively, as when using crutches or gymnastically supporting the body on parallel bars.

(c) Protraction of the scapula

Movement produced by:
• serratus anterior
• pectoralis minor
• latissimus dorsi (a 'strapping'effect by upper fibres)

(d) Retraction of the scapula

Movement produced by:
• trapezius (middle/lower fibres)
• rhomboids

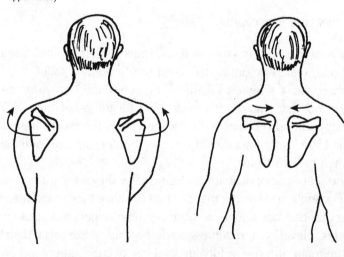

These movements are produced either when punching and pushing (protraction) or when bracing the shoulders back (retraction).

(e) Forward rotation of the scapula

Movement produced by:
- trapezius (upper fibres)
- serratus anterior (lower fibres)

(f) Backward rotation of the scapula

Movement produced by:
- gravity
- graded relaxation of trapezius (upper fibres)and serratus anterior (lower fibres)
- levator scapulae
- rhomboids
- pectoralis minor
- trapezius (lower fibres)

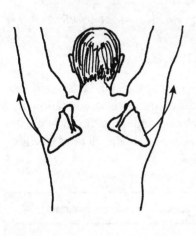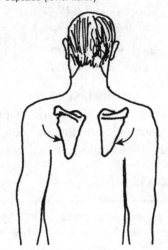

The forward rotation of the scapula turns the glenoid cavity upwards and allows the arm to be raised above the head. Backward rotation is usually a return to the anatomical position by gradual relaxation of the forward rotator muscles, but can be produced actively.

Figure 10.5 Movements of the scapula

Note that in reality the scapular movements described above may to some extent combine. For example, retraction and backward rotation, and protraction and forward rotation may occur together.

Scapulohumeral movements

If the arms are held loosely at the sides (the position of rest), the stability of the glenohumeral joint is maintained by the superior part of the joint capsule, by the rotator cuff muscles that blend with it and by the support of the glenohumeral and coracohumeral ligaments. In this position the inferior portion of the capsule hangs down loosely as a dependent pouch (Figure 10.2(a)). This laxity allows the freedom of movement which the shoulder (and hand) needs.

Study tasks

- Examine all movements available at the glenohumeral joint, and the ways in which the scapula enhances the scope of these movements. (Before proceeding with this task, glance ahead at the series of diagrams that accompany the section entitled 'Total shoulder girdle movements', making sure that you understand the terms 'flexion', 'extension', 'abduction', 'adduction', 'medial and lateral rotation', and 'circumduction', as applied to the shoulder).
- Ask a colleague to slowly perform the shoulder movements listed in Table 10.1. As they do so, observe and palpate the movements of the scapula using the classification shown in Figure 10.5. Record the type of scapular movement that occurs with reference to Table 10.1. Note that it may be helpful to gently grip the inferior angle of the scapula in order to detect movement. Do not restrain the scapula too much, since this will inhibit glenohumeral movement.
- Repeat the movements, but this time estimate the range of movement (in degrees) through which the glenohumeral joint moves before the scapula starts to move. Test a number of colleagues if possible, and record your findings with reference to Table 10.2. Your observations of total shoulder movement should provide a realistic appreciation of what Codman (1934) called 'scapulohumeral rhythm'.

Clinical Note

If the capsule becomes inflamed, it may thicken and fibrose, accompanied by a synovitis that results in contracture and adhesions, giving rise to the condition known as 'adhesive capsulitis' (frozen shoulder). The laxity of the capsule, which is an asset for movement, now becomes a liability by providing a greater surface area for adhesions.

In the position of rest, the joint capsule has a forward and medial twist. This helps to produce the necessary rotation that accompanies abduction (see below), and prevents the greater tuberosity of the humerus from impinging on the glenoid labrum.

Table 10.1

Glenohumeral movement	Type of accompanying scapular movement
Flexion	
Extension	
Abduction	
Adduction	
Medial rotation	
Lateral rotation	

Table 10.2

Glenohumeral movement	Angle at whch scapular movement commences (DEGREES)
Flexion	
Extension	
Abduction	
Adduction	
Medial rotation	
Lateral rotation	

A useful working ratio of glenohumeral:scapular movement is 2:1 (Williams, 1995). Thus, for every 3° of movement in the shoulder girdle, the glenohumeral joint contributes 2° and the scapula 1°. Readers may encounter some variation in the literature concerning the exactitude of this point.

The role of the clavicle

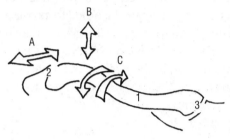

1 The clavicle
2 Acromioclavicular joint (synovial: plane) – allowing permissive movement when the scapula moves
3 Sternoclavicular joint (synovial: saddle) acts as a pivot to allow all available movements of the clavicle

Movements
A Protraction/retraction
B Elevation/depression
C Rotation

Figure 10.6 Schematic illustration of the right clavicle: joints and movements (anterior view)

The clavicle is in effect the strut that attaches the shoulder (part of the appendicular skeleton) to the body (axial skeleton), and the sternoclavicular joint is the only articulation that attaches the shoulder to the axial skeleton. The clavicle also rotates and acts with the scapula to increase the range of total shoulder movement, and its S-shaped crank is important in this respect. For this reason the functional importance of the clavicle, and the articulating joints at each end, should not be overlooked when assessing the range of movement available in the entire shoulder girdle.

Total shoulder girdle movements

The conventional description of joint movement which has been used in this chapter so far (and will continue to be used) refers to axes of movement or degrees of freedom. Thus, movement through one degree of freedom allows flexion/extension through the transverse axis; a second degree of freedom allows abduction/adduction through the anteroposterior axis; and a third degree of freedom allows rotation movements through a longitudinal axis.

However, this is a generalized description that can be applied to all joints as appropriate, and is not really adequate if a more detailed appreciation of joint mechanics is required. Inspection of

Study tasks

- Try to obtain several clavicle bone specimens and note the following features: (i) the S-shape is quite variable; (ii) the right clavicle is often thicker; (iii) the clavicle of a female may be smaller.
- Palpate the differences between right and left clavicles on several colleagues of both sexes.
- Highlight the details shown in Figure 10.6.
- Examine each end of a clavicle and consider how the differences in structure reflect the differences in function of each joint. Refer also to the details of the clavicular joints given in Figure 10.2(a).

the glenoid surface of the scapula reveals a shallow contact surface for the humerus, allowing a wide range of movement with minimal bone-to-bone contact. Approximately only one-third of the humeral head is in contact with the glenoid surface at any given time, and when movement takes place there is essentially a combination of slide (translation), spin and roll at the joint surfaces, which is necessary for the relatively large humeral head to travel over the small glenoid cavity, and to avoid dislocation. An appreciation of these arthrokinematic concepts is important in order to provide effective manipulative treatment for patients with glenohumeral restriction. These terms have been defined and explained in Chapter 3, and will be used as appropriate in the ensuing description of total shoulder girdle movements, particularly with reference to the behaviour of the glenohumeral joint.

Flexion

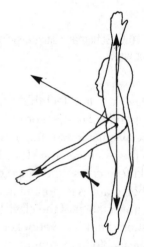

Phase 3: 120–180°
Movement produced by:
Phase 1 and 2 muscles, plus muscles that produce vertebral sidebending (contralateral side)

Phase 2: 60–120°
Movement produced by:
Phase 1 muscles, plus muscles that forwardly rotate the scapula:
- trapezius (upper fibres)
- serratus anterior (lower fibres)

Phase 1: 0–60°
Movement produced by:
- pectoralis major (clavicular part)
- anterior deltoid
- coracobrachialis assisted by:
- biceps brachii

Final phase of flexion (or abduction) of the shoulder girdle: 120–180°

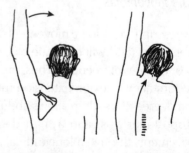

Note: The rotator cuff muscles (Figure 10.4) also provide support as the glenohumeral ligaments slacken when the arm is elevated.

Figure 10.7 Flexion of the shoulder

The reason for the tendency of the hands to converge is that the plane of the glenoid surface of the scapula which lies against the thoracic cage forms an angle of approximately 30° with the midline. In order to compensate for this the humeral head is retroverted by a similar angle of 30–40° (compared with the femur, which has an angle of anteversion; see p. 152). Like the femur the humerus also has an angle of inclination.

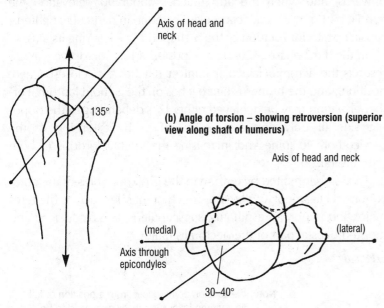

(a) The angle of inclination (anterior view)

Axis of head and neck

135°

(b) Angle of torsion – showing retroversion (superior view along shaft of humerus)

Axis of head and neck

(medial) (lateral)

Axis through epicondyles

30–40°

Figure 10.8 The humerus: angles of inclination and torsion

With the arms hanging loosely at the sides, the stability of the glenohumeral joint is maintained by tension in the superior capsule, the glenohumeral and coracohumeral ligaments and the rotator cuff muscles. Both the glenoid and humeral articular surfaces are lined with hyaline cartilage which is thickest at the periphery of the glenoid fossa and at the centre of the humeral head. The articular surfaces may therefore act as a kind of suction cup when compressed under load, but the laxity of the capsule means that the head of the humerus can be distracted (moved away from the socket) when the shoulder is relaxed. In the first phase of flexion (0–60°) the movement at the humeral head appears to involve spin through a transverse axis. However, the tendency for the arm to drift towards the midline is functionally avoided by the introduction

Study task

- Palpate posterior rotation in your own clavicle as you flex your shoulder from 0° to 180°, by placing the palmar surface of three fingers over the clavicle near the sternoclavicular joint.

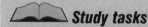
Study tasks

- Highlight the details shown.
- Muscle check.
- Swing your arms loosely back into extension and note how they tend to diverge away from the body. This is the converse of flexion but is caused by the same angulation of the glenoid cavity.
- Repeat the movement, but keep your arms straight, parallel and equidistant as they move into about 50° extension. Analyse the individual movements that are taking place.
- Consider why the range of movement is apparently much less than in flexion.
- Observe the scapular movements that occur, on a colleague, by gently grasping and palpating the movement of the inferior angle of the scapula.
- Also palpate the individual movements of the clavicle and sternoclavicular and acromioclavicular joints during extension.

of an element of lateral rotation and abduction. This introduces elements of roll and slide at the joint surfaces.

Once the arm begins to elevate, the superior structures begin to slacken, which renders the shoulder unstable. This is counteracted by the opposing pull of the rotator cuff muscles, and the commencement of forward rotation of the scapula which ensures that contact is maintained between the glenohumeral joint surfaces.

The second phase of flexion (60–120°) has now begun. Forward rotation of the scapula allows the glenoid fossa to face upwards, and when the arm reaches about 90° elevation the conoid part of the coracoclavicular ligament, in particular, tightens, causing posterior rotation of the S-shaped clavicle along its axis.

In the third phase (120–180°), posterior rotation of the clavicle restricts the acromioclavicular joint (it reaches the 'close-packed' position), and the humerus encroaches on the glenoid labrum. Further elevation may be achieved either by sidebending to the opposite side (if only one arm is raised) or by extension of the cervicothoracic spine and increasing the lumbar lordosis (if both arms are raised).

Flexion is inhibited by tension in the posterior fibres of the coracohumeral ligament and by the stretch of muscles such as the teres major and minor, infraspinatus, subscapularis, latissimus dorsi and the sternocostal part of the pectoralis major.

Extension

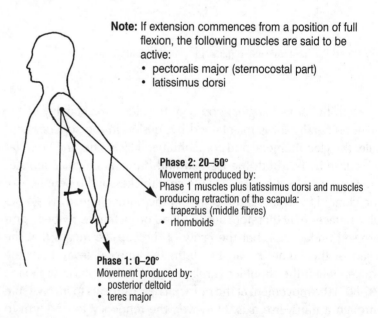

Note: If extension commences from a position of full flexion, the following muscles are said to be active:
- pectoralis major (sternocostal part)
- latissimus dorsi

Phase 2: 20–50°
Movement produced by:
Phase 1 muscles plus latissimus dorsi and muscles producing retraction of the scapula:
- trapezius (middle fibres)
- rhomboids

Phase 1: 0–20°
Movement produced by:
- posterior deltoid
- teres major

Figure 10.9 Extension of the shoulder

In order to appreciate the reduced degree of extension available in the standing position, reflect on the evolutionary implications of a change from a partially quadrupedal to a bipedal stance. If you fully extend your arms in the standing position, but then flex your lumbar spine and hips, you will almost assume a diving pose. In other words, in the quadrupedal position you have overextended and as such would not require the same degree of extension that is needed in a bipedal stance. It may now be clear that most extension movements in life are usually relative extension from varying degrees of flexion. This may involve forceful use of the extensor muscles, or may involve a gradual relaxation of the flexor muscles under the influence of gravity. It depends on the circumstances.

Extension will also involve varying degrees of medial rotation in order to compensate for the angle of the plane of the glenoid surface of the scapula. As in the case of flexion, it is a rather impure compound movement, and will also involve elements of spin, slide and roll.

Extension is limited by tension in the flexor muscles and ultimately by tension in the anterior capsule, the anterior part of the coracohumeral and glenohumeral ligaments, and by the palpable limit of retraction in both the scapula and clavicle.

Abduction

Phase 3: 150–180°
Movement produced by:
Phase 1 and 2 muscles plus muscles that produce vertebral sidebending (contralateral side)

Phase 2: 90–150°
Movement produced by:
Phase 1 muscles plus muscles that forwardly rotate the scapula:
• trapezius (upper fibres)
• serratus anterior (lower fibres)

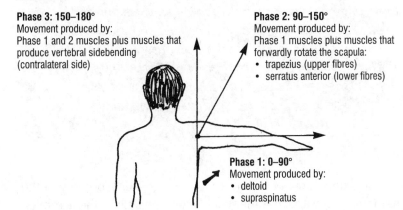

Phase 1: 0–90°
Movement produced by:
• deltoid
• supraspinatus

Figure 10.10 Abduction of the shoulder

This is a movement of great importance and clinical significance which illustrates the strengths and weaknesses of the shoulder girdle. Some authorities classify flexion and abduction under the general heading of 'elevation' of the shoulder, since their final position is the same (phase 3 in Figures 10.7 and 10.10).

Study tasks

• Highlight the details shown.
• Muscle check.
• Observe a healthy colleague with a normal shoulder performing abduction, and pay special attention to what happens at the glenohumeral joint if elevation continues to the point where the hands meet overhead.
• Repeat the movement with 'rounded shoulders'. Consider whether this makes the movement easier or more difficult, and explain your findings in anatomical terms.
• Observe the movement of the scapula during abduction, and also palpate the movements that take place at the sternoclavicular and acromioclavicular joints and along the shaft of the clavicle.

Abduction is initiated by the contraction of deltoid and supraspinatus muscles, with the latter also producing a compressive force to the glenohumeral joint surfaces. As abduction continues, the remaining rotator cuff muscles act to depress the humeral head, thus counteracting the upward force of the deltoid. While the deltoid tends to produce an upward roll of the humeral head, which encourages an upward migration, the rotator cuff muscles (e.g. the subscapularis) encourage a downward slide, and therefore stability is maintained.

In the first phase of abduction (0–90°), movement occurs in the glenohumeral joint with only limited movement of the scapula or clavicle, at least in the early stages. This is because of the fact that the muscles that initiate the movement (the supraspinatus and deltoid) originate on the scapula/clavicle, which must remain stable to allow their insertion into the humerus to produce movement.

As the movement proceeds, the superior part of the glenohumeral ligament slackens, while the middle and inferior fibres tighten.

Clinical Note

Dislocation of the glenohumeral joint will occur if a strong downward force is applied to the semi-abducted shoulder. This may occur in contact sports when a tackle is being discouraged by an outstretched arm. The problem may become recurrent if the capsule and ligaments are weakened, and in this context the strength of the rotator cuff muscles assumes paramount importance.

As abduction continues beyond 90°, the greater tuberosity of the humerus begins to close the space under the arch of the acromion and coraco-acromial ligament (the subacromial surface) (Figure 10.11). The medial twist in the joint capsule produces sufficient tension to cause the humerus to rotate laterally so that the greater tuberosity avoids impinging upon the acromial arch and glenoid labrum. Lateral rotation tightens all three bands of the glenohumeral ligament of the humerus and also changes the relative attachment points of the biceps brachii, allowing it to play a role in further abduction movements (and thus provide added anterior support).

Full lateral rotation with abduction is regarded as the 'close-packed' position for the shoulder (Williams, 1995).

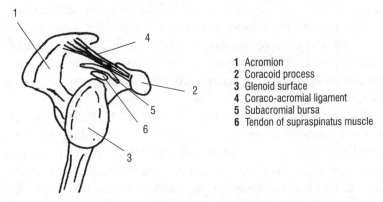

1 Acromion
2 Coracoid process
3 Glenoid surface
4 Coraco-acromial ligament
5 Subacromial bursa
6 Tendon of supraspinatus muscle

Figure 10.11 Subacromial surface of the scapula (lateral view)

Clinical Note

The subacromial surface forms a tunnel occupied by the tendon of supraspinatus and the overlying subacromial bursa which protects the tendon from the acromion process and coracoacromial ligament. If either the tendon or the bursa is inflamed (or both), abduction becomes painful in an arc of movement classically between about 45° and 160° as the greater tuberosity compresses the inflamed subacromial structures. This is known as the 'painful arc' and is a useful sign in clinical diagnosis.

   In the second phase of abduction (90–150°), the scapula is now rotating in a forward direction (activated by the muscle fibres of the upper trapezius and lower serratus anterior), which has the effect of elevating the glenoid cavity and also increases the mechanical efficiency of the deltoid muscle by allowing it to maintain its resting length (Cailliet, 1991b). As the scapula rotates, the movement is transmitted via the acromioclavicular joint, which does not move greatly due to its planar surfaces, through to the sternoclavicular joint, which is saddle shaped and more versatile. The clavicle elevates, but increasing scapular rotation causes the conoid part of the coracoclavicular ligament, in particular, to tighten. It therefore pulls the clavicle into backward rotation, as in flexion.

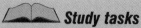 *Study tasks*

- Consider how an exaggerated thoracic kyphosis might hinder abduction of the shoulder.
- If a patient experiences restrictions in abduction, consider what other movements at the glenohumeral joint surface you might wish to examine in order to restore proper function.

 *Study tasks*

- Highlight the details shown in Figure 10.12.
- Muscle check.
- Ask a colleague to perform adduction from the anatomical position with flexion and then extension of the shoulder, and then from a position of abduction. Note any rotation movements, and observe and palpate accompanying movements at the scapula and clavicle and associated joints.
- Try to discover which structures limit movement at the end of range.

The acromioclavicular joint reaches a tight close-packed position, and abduction can continue up to about 150° by means of scapular rotation and rotation at the sternoclavicular joint, which are said to contribute about 60° of this movement (Kapandji, 1982). By now, the middle and inferior fibres of the glenohumeral ligament and capsule are taut; and if the joint is laterally rotated and fully abducted the shoulder is in its close-packed position. The greater tuberosity of the humerus will be in contact with the glenoid labrum.

In the third phase, abduction can reach 180°, but this requires extra movements of extension through the vertebral column (and specifically the cervicothoracic and upper thoracic spine) if shoulder elevation is bilateral, or of sidebending if the movement is unilateral. This is the same endpoint as in final flexion (Figure 10.7).

Adduction

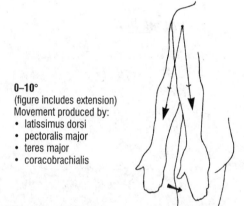

0–10°
(figure includes extension)
Movement produced by:
- latissimus dorsi
- pectoralis major
- teres major
- coracobrachialis

Figure 10.12 Adduction of the shoulder (posterior view)

Pure adduction is not possible because in the anatomical position the trunk obstructs the movement. In order to avoid this the shoulder must be moved initially into either flexion or extension, usually accompanied by varying degrees of medial or lateral rotation. In practice, the adductor muscles are usually utilised from a position of abduction, and the movement might be properly termed 'relative' adduction. In any case it is difficult to generalise about adduction movements at the shoulder, and it is good practice to carefully analyse what is happening in any given situation.

Medial and lateral rotation

The rotatory movements permitted by the spheroidal shape of the glenohumeral joint cannot be described as pure spin unless the arm is held in approximately 90° abduction. Only at this point do the long axis and mechanical axis of the joint coincide. In all other positions, rotation along the long axis involves roll and slide. In addition, rotatory movements at the shoulder are rarely performed with a straight arm. Normally the elbow is flexed, and in the case of medial rotation the movement is obstructed by the trunk. Medial and lateral rotation are also often performed with an abducted shoulder.

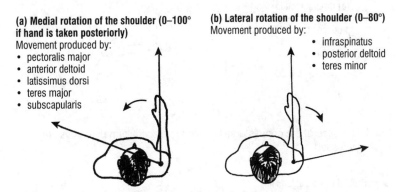

(a) Medial rotation of the shoulder (0–100° if hand is taken posteriorly)
Movement produced by:
- pectoralis major
- anterior deltoid
- latissimus dorsi
- teres major
- subscapularis

(b) Lateral rotation of the shoulder (0–80°)
Movement produced by:
- infraspinatus
- posterior deltoid
- teres minor

Figure 10.13 Rotation movements of the shoulder

Study tasks
- Highlight the details shown in Figure 10.13.
- Observe and palpate a colleague performing the movements and analyse the accompanying movements of the scapula and clavicle and associated joints.

You should have noticed that medial rotation is facilitated by protraction of the scapula. It is functionally important that the hand can be moved behind the body to the back, for example when a woman unfastens her own bra strap. This increases the range of medial rotation and represents medial rotation with extension. The end of range position is reached as the clavicle moves forwards at the acromioclavicular joint, accompanied by a backward glide or retraction of the sternoclavicular joint. These movements are checked by the anterior sternoclavicular ligament and the posterior fibres of the costoclavicular ligament, by tension in the trapezoid part of the coracoclavicular ligament, by the acromioclavicular joint capsule, and in extreme medial rotation (the so-called 'half Nelson' position) by stretch of the lateral rotator muscles. The main ligaments of the glenohumeral joint are relaxed in medial rotation.

Lateral rotation eventually involves some retraction or backward rotation of the scapula. The clavicle moves backwards or retracts, and the movement is checked by the anterior sternoclavicular ligament, the anterior fibres of the costoclavicular ligament, and the glenohumeral ligament and anterior fibres of the coracohumeral ligament. Tension in the antagonistic muscles is probably less significant than in extreme medial rotation.

Circumduction and horizontal flexion/extension

Circumduction is a movement that combines all of the components previously described, utilising the three degrees of freedom available in a multi-axial spheroidal joint. Horizontal flexion and extension are other combined movements that may be described, and consist of shoulder movements in a transverse plane with the shoulder in a position of 90° abduction.

Study tasks

- Using your knowledge of shoulder movements, analyse the individual components that make up the movement of circumduction.
- Do the same for horizontal flexion and extension, and in this case analyse the movements at the glenohumeral joint surface using the concepts of slide, roll, spin and swing.

(a) Circumduction
Either clockwise or anticlockwise, produced by muscles of:
- flexion
- abduction
- extension
- adduction

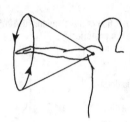

Note: The base of the cone is greatly enlarged if scapular movements are increased.

(a) Horizontal flexion/extension
(variable range through longitudinal axis)

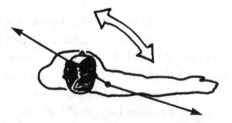

Figure 10.14 Circumduction and horizontal movements of the shoulder

Chapter 11

The elbow

Introduction

The functional role of the upper limb is to allow the hand to move freely in space. The elbow joints, which link the upper arm and forearm, allow adjustments to be made in height and length. The specialised rotatory movements of pronation and supination also allow the hand to be placed in a functionally optimum position once the spheroidal joint of the shoulder has chosen the general direction of movement.

Three joints may be identified. Two are between the distal end of the humerus and the proximal ulna and radius, respectively; these joints are concerned with flexion/extension movements. The third joint is the articulation between the proximal ends of the radius and ulna, and allows the specialised rotatory action of pronation and supination.

The bones

The main hinge joint that allows flexion and extension is formed by the trochlear (pulley-shaped) surface of the distal humerus (the trochlea), which articulates with the trochlear notch of the ulna and is supported anteriorly by the prominent ledge of the coronoid process on the ulna. This is the humero-ulnar joint. The other joint that participates in flexion/extension is that between the spheroidal capitulum on the humerus and the concave superior surface of the radius. This is the humeroradial joint, which looks rather like a spheroidal joint but is prevented from being so by the proximity of the humero-ulnar surfaces. In fact this joint is subordinate to the

- Highlight the bone features shown, and identify the features on bone specimens.
- Palpate as many of the bony features as you can on your own elbow by supporting it in semi-flexion with your palpating hand. Place your thumb on the prominent medial epicondyle. Slide your middle finger under and across to the less prominent lateral epicondyle. Just above lies the lateral supracondylar ridge, and just below lies the head of the radius, which is palpable in the movements of pronation and supination. Now move your index finger on to the olecranon and, keeping thumb on the medial epicondyle and middle finger on the lateral epicondyle, fully flex your elbow. Then extend it and notice how the relative position of the olecranon changes*.
- Identify these features on a colleague.
- Rest your semi-flexed elbow on a table and palpate the olecranon fossa just above the olecranon. The tendon of the triceps which attaches to the olecranon can be easily felt if a rapid slight extension movement is made.

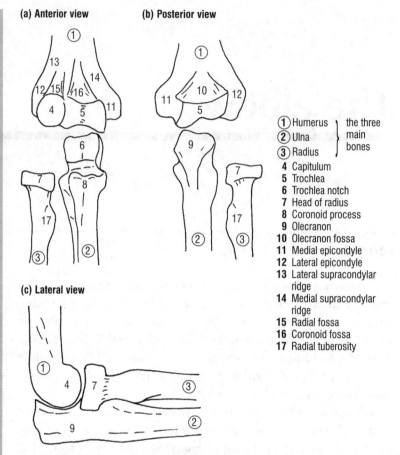

Figure 11.1 The bones of the elbow

(a) Anterior view (b) Posterior view (c) Lateral view

1 Humerus
2 Ulna
3 Radius
} the three main bones

4 Capitulum
5 Trochlea
6 Trochlea notch
7 Head of radius
8 Coronoid process
9 Olecranon
10 Olecranon fossa
11 Medial epicondyle
12 Lateral epicondyle
13 Lateral supracondylar ridge
14 Medial supracondylar ridge
15 Radial fossa
16 Coronoid fossa
17 Radial tuberosity

extent that surgical removal of the head of the radius does not greatly inhibit flexion/extension movements of the elbow.

Posterior stability comes from the supporting ledge of the olecranon process of the ulna. The prominence of the coronoid and olecranon processes, along with the disc-shaped head of the radius mean that in flexion approximation with the humerus would be prevented without the presence of the coronoid, olecranon and radial fossae on the opposing surfaces of the humerus (Figure 11.1(a–b)). The interlocking nature of the bones allows the elbow to combine stability with versatility.

Laterally and medially on the humerus, the epicondyles and supracondylar ridges from important muscle attachment points.

The relative change in position of the olecranon noticed in the Study Task* is due to the articular surface of the trochlea on the humerus, which extends further distally than the capitulum. Notice that the medial lip or flange of the trochlea is longer than the lateral lip. The effect of this increases as the trochlear surface on the ulna engages the trochlea when the elbow moves into extension, and the shaft of the ulna deviates away from the midline, as represented by the shaft of the humerus.

This angulation, which is most apparent in full extension of the elbow, and less so in flexion, is known as the 'carrying angle' and is an example of a 'valgus' angulation, since it deviates from the midline. It becomes more apparent in hyperextension, which is quite common, and in females, perhaps due to greater ligamentous laxity. There is a noted difference between the average carrying angle of males and females (see Figure 11.2(a)) and care must be taken to clarify which angle is being measured when describing abnormal increases or decreases in the carrying angle, which are perhaps best referred to as 'cubitus valgus' and 'cubitus varus'.

(a) The carrying angle　　　　**(b) Cubitus valgus/varus**

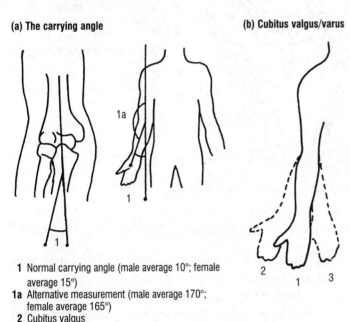

1 Normal carrying angle (male average 10°; female average 15°)
1a Alternative measurement (male average 170°; female average 165°)
2 Cubitus valgus
3 Cubitus varus

Figure 11.2 Aspects of the carrying angle

The joints and ligaments

The three joints of the elbow are enclosed by a common capsule which is broad and thin. Anteriorly, it is attached superiorly on the humerus at epicondyle level. Inferiorly, it is attached to the coronoid process on the ulna and the annular ligament on the radius (see below). Posteriorly, it attaches above the capitulum and

📖 *Study tasks*

- Highlight the features shown in Figure 11.3.

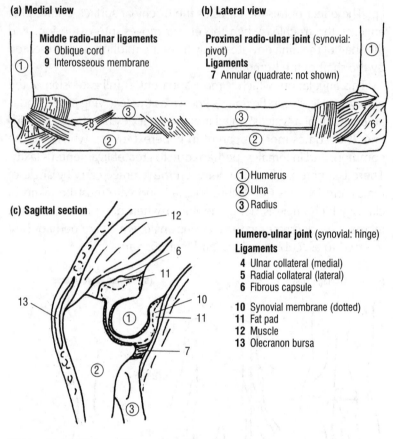

(a) Medial view

Middle radio-ulnar ligaments
8 Oblique cord
9 Interosseous membrane

(b) Lateral view

Proximal radio-ulnar joint (synovial: pivot)
Ligaments
7 Annular (quadrate: not shown)

(c) Sagittal section

① Humerus
② Ulna
③ Radius

Humero-ulnar joint (synovial: hinge)
Ligaments
4 Ulnar collateral (medial)
5 Radial collateral (lateral)
6 Fibrous capsule

10 Synovial membrane (dotted)
11 Fat pad
12 Muscle
13 Olecranon bursa

Figure 11.3 Joints and ligaments of the elbow

Clinical Note

Inflammation of the olecranon bursa which overlies the bony prominence of the olecranon gives rise to a swelling known as 'olecranon bursitis'. It is often caused by constant pressure or the repetitive impact of the elbows on a hard surface such as a table or desk, and for this reason is sometimes referred to as 'student's elbow'.

trochlea, and to the olecranon and proximal radio-ulnar joint, beneath the annular ligament.

The synovial membrane lines the capsule and annular ligament. Posteriorly it has a fold which partly separates the humero-ulnar and humeroradial joints.

The presence of the various fossae and the gaps between the articulations also gives rise to a number of vascular and well inner-vated (hence pain-sensitive) fat pads. The largest one lies in the ole-cranon fossa (Figure 11.3(c)). Others are found in the coronoid and radial fossae, on each side of the trochlear notch and in the poste-rior fold between the two articulations. Bursae are also present, of which the largest and most significant is the olecranon bursa over-lying the olecranon process (Figure 11.3(c)). The capsule is rein-forced by the various ligaments, whose main details are shown in Figure 11.3(a–b).

The carrying angle imposes special demands on the medial (ulnar) collateral ligament, which is designed to prevent valgus strain, especially the anterior part, which is taut in extension. The posterior element is taut in flexion. It is perhaps worth noting that the most powerful usage of the elbow, as in throwing, supporting weight and so on, tends to require stability on the medial side. The lateral (radial collateral) ligaments are less strong, since they attach to the lateral epicondyle and pass to the annular ligament only. This does not seem to pose a major problem, since varus forces are less common at the elbow. Pauly *et al.* (1967) have suggested that one of the roles of the anconeus muscle (an apparently weak elbow extensor) is to provide additional support against varus strain.

The annular ligament, which is lined by articular hyaline cartilage and a reflection of synovial membrane, surrounds the head of the radius. It is supported inferiorly by the membranous 'quadrate' ligament, which tightens at the end range of pronation and supination (see p. 124).

Study tasks

- Rest your elbow in a relaxed position of flexion and palpate the structures around the medial epicondyle. The medial (ulnar) collateral ligament cannot be readily felt but its position and attachments should be identified. Gently find the cordlike ulnar nerve which is mobile and palpable in the sulcus (hollow) between the epicondyle and olecranon. Sharp pressure will reproduce the unpleasant 'funny bone' sensation of tingling in the distribution of the ulnar nerve.
- Palpate the lateral epicondyle and the head of the radius and annular ligament. The lateral radial collateral ligament is not easily palpable, but its position should be identified.

Clinical Note

The complexity and mobility of the elbow joints means that the major nerves that pass down the arm may become entrapped. In the case of the ulnar nerve an entrapment neuropathy may affect the nerve as it passes around the medial epicondyle or as it pierces the intermuscular septum close to the medial head of the triceps muscle.

Muscles and movements

The defining shape of the surface anatomy of the elbow reflects the powerful attachment points of the wrist flexor muscles (to the medial epicondyle), of the wrist extensors (to the lateral epicondyle), of the elbow flexor muscles anteriorly and of the elbow extensors posteriorly. It is likely that the muscles that produce the specialised movements of pronation and supination at the elbow (Figure 11.7) are also important stabilisers when the elbow is under strain.

(a) Lateral epicondyle* – site of common extensor tendon (of wrist)

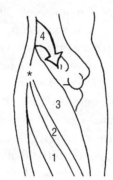

1 Extensor carpi ulnaris
2 Extensor digitorum
3 Extensor carpi radialis brevis
4 Site of extensor carpi radialis longus (schematic representation only)

(b) Medial epicondyle* – site of common flexor tendon (of wrist)

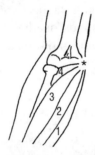

1 Flexor carpi ulnaris
2 Palmaris longus
3 Flexor carpi radialis
4 Pronator teres

Figure 11.4 The medial and lateral epicondyles of the humerus

Clinical Note

Repetitive strain injuries of the common flexor and extensor tendons of the wrist may give rise to chronic inflammation at the elbow which is known as lateral epicondylitis ('tennis elbow') if the extensor tendon is affected or medial epicondylitis ('golfer's elbow') if the flexor tendon is involved.

Flexion

0–145°
Movement produced by:
- brachialis
- biceps brachii
- brachioradialis
assisted by:
- pronator teres

Figure 11.5 Flexion of the elbow

The brachialis is regarded as the prime flexor muscle of the humero-ulnar joint due to its insertion directly into the ulna. The biceps brachii, which inserts into the radial tuberosity, is an important muscle of supination and is also important as an elbow flexor in the supinated or neutral position (Figure 11.7) (Basmajian and Latif, 1957). Larson (1969) has found that maximal flexion force is generated when the elbow is supinated or in neutral, and least in a position of pronation. However, as Currier (1972) has pointed out, much of the work on muscle power is based on electromyographic analysis, which may not be a reliable indicator of maximum muscle force.

London (1981) has suggested that the movement at the humero-ulnar joint surfaces during flexion/extension is mainly one of slide changing to roll in the final 5–10°. (Compare this with the knee; see p. 176, 'Movements at the joint surfaces'.)

Extension

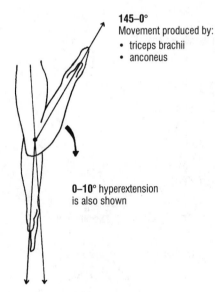

145–0°
Movement produced by:
• triceps brachii
• anconeus

0–10° hyperextension
is also shown

Figure 11.6 Extension of the elbow

The elbow is extended in the anatomical position, and so active extension commences from various positions of flexion. Basmajian (1969) concludes that the medial head of the triceps brachii is the most powerful muscle of extension. Pauly *et al.* (1967) regard the anconeus as an important stabiliser, particularly against varus forces.

 Study tasks

- Highlight the details shown in Figure 11.5.
- Muscle check.
- Palpate the biceps brachii tendon and aponeurosis by making a fist and palpating your flexed elbow. The tendon may be palpated as it crosses the flexed elbow. The aponeurosis forms a sharp edge palpable on the medial border of the tendon.
- Consider which tissues limit the range of flexion.
- Palpate the brachioradialis muscle by placing your fist in a position midway between pronation and supination, under a table. Exert gentle effort as if to lift the table, and the contracting muscle belly will be seen just distal to the elbow.

 Study tasks

- Highlight the details shown in Figure 11.6.
- Muscle check.
- Palpate the bony limit of full extension by supporting your elbow with one hand and gently allowing your forearm to drop into extension.
- Assess how much hyperextension is present in your own and colleagues' elbows in full passive extension.

Full extension is reached when the olecranon process on the ulna meets the olecranon fossa on the humerus. It may be possible to extract as much as 10–15° further extension if a colleague's elbow is gently taken into passive extension, especially if there is laxity present in the ligaments and capsule. This is a more common finding in females.

Pronation and supination

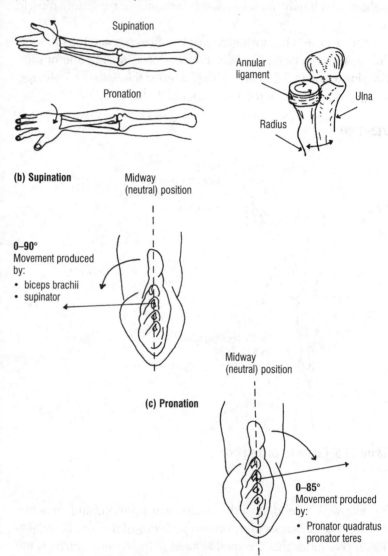

(a) The proximal radio-ulnar joint

Supination

Pronation

Annular ligament

Radius

Ulna

(b) Supination

Midway (neutral) position

0–90°
Movement produced by:
• biceps brachii
• supinator

(c) Pronation

Midway (neutral) position

0–85°
Movement produced by:
• Pronator quadratus
• pronator teres

Figure 11.7 The proximal radio-ulnar joint and the movements of pronation and supination

Pronation and supination movements allow the hand to rotate, but the axis of rotation is diagonal, and limited by the extent to which the radius can rotate across the ulna. Basmajian (1969) has demonstrated that the pronator quadratus is the most powerful muscle of pronation, and also that the pronator teres is the only other muscle involved in the movement.

Supination is the more powerful movement due to the involvement of the biceps brachii, as well as the supinator. Basmajian (1969) discounts any contribution from the brachioradialis, which had previously been regarded as an assistor muscle from a position of extreme pronation.

Chapter 12

The wrist

Introduction

The role of the wrist is to facilitate the most effective positioning of the hand. In this respect it is able to provide a stable platform for the most efficient and powerful use of the flexor and extensor muscles that originate on the bones of the forearm and which activate the hand in its various positions of prehensile activity. At other times the carpal bones may possess sufficient mobility to allow the wrist greater freedom of movement, which is important during delicate hand movements.

The bones

1–5 Metacarpal bones of the hand
6 Trapezium ⎫ Distal
7 Trapezoid ⎬ carpal
8 Capitate ⎪ row
9 Hamate ⎭
10 Scaphoid ⎫ Proximal
11 Lunate ⎬ carpal
12 Triquetral ⎪ row
13 Pisiform ⎭
14 Radius
15 Ulna

Joints
- radiocarpal (ellipsoid)
- midcarpal (compound saddle)
- intercarpal (complex plane)
- carpometacarpal (1st = saddle 2–5 = plane)
- intermetacarpal (plane)

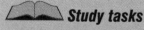

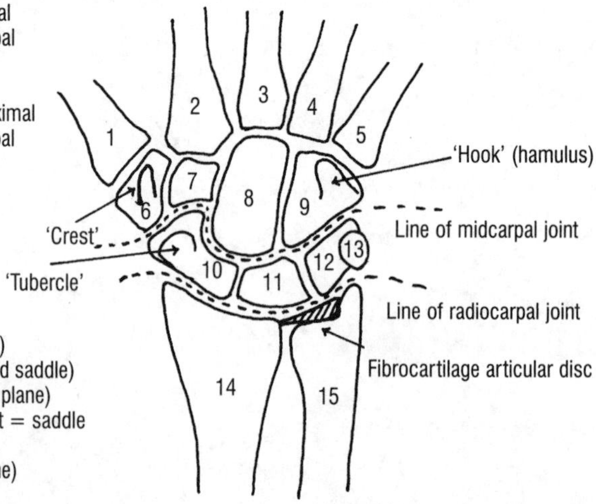

Figure 12.1 The bones and joints of the wrist (palmar surface)

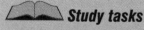

Study tasks

- Shade the radius, ulna and the proximal and distal rows of carpal bones in separate colours.
- Highlight the names of the bones and the details given.
- Obtain bone specimens and examine the features of the distal radius and ulna, and each carpal bone in turn. Pay particular attention to the following: (i) the styloid processes of the radius and ulna; (ii) the scaphoid tubercle; (iii) the crest of the trapezium; (iv) the hook of the hamate.

The wrist may be described as a multi-articular complex of bones, which is conventionally divided into two rows, and which also defines the functional articular surfaces.

There is a proximal arrangement of four individual bones that articulate with the distal radius but are separated from the ulna by a triangular fibrocartilage disc which cushions compressive loading of the wrist. This joint 'line' is referred to as the radiocarpal joint.

The proximal row articulates with a distal row of four bones forming the midcarpal joint; and this distal row in turn articulates with the proximal bases of the metacarpals of the hand.

The joints and ligaments

The entire array of carpal bones functions as a multi-articular complex, held together by a complicated network of ligaments and traversed by palmar and dorsal muscle tendons. These muscle tendons are held in position by the flexor retinaculum on the palmar side and the extensor retinaculum on the dorsal side. These are retaining bands of deep fibrous fascia, of which the flexor retinaculum is of greater clinical significance. With attachments to the scaphoid tubercle, crest of the trapezium, hook of the hamate, and the pisiform, the flexor retinaculum forms the osseofibrous 'carpal tunnel' (Figure 12.2(a) and (d)).

The extensor retinaculum passes obliquely across the dorsal surface of the wrist from the triquetral and pisiform to the anterior surface of the radius, and holds the extensor tendons in place.

The main joints usually identified in the wrist are all compound synovial joints. They are the radiocarpal joint, the midcarpal joint and the general assemblage of intercarpal joints. For completeness, mention should also be made of the distal radio-ulnar joint, where the rotatory action of pronation/supination which takes place at the elbow is also represented at the wrist.

Clinical Note

The median nerve, which is sensory to the thumb and index and middle fingers, passes through the carpal tunnel and may become compressed as a result either of inflammation of structures such as the flexor tendons or carpal joints; or of fluid retention during pregnancy or the menopause. The most common symptom of 'carpal tunnel syndrome' is nocturnal tingling and numbness in the lateral three digits of the hand.

Study tasks

- Consider why each bone was given its particular name.
- Identify each bone in your own wrist as follows:
  - *Scaphoid*: Extend your wrist and palpate its tubercle. With index finger on this, place your thumb on the scaphoid in the anatomical 'snuffbox' (see p. 140). You will now be gripping your scaphoid, and some movement will be palpable with your wrist in neutral.
  - *Lunate*: Slide your palpating thumb from the scaphoid to the adjacent lunate with the wrist in slight flexion. Just proximal lies the dorsal tubercle of the radius.
  - *Triquetral*: This lies palpably below the *pisiform*, which is prominent and immobile in full wrist extension; both bones may be easily grasped and moved if the wrist is flexed and adducted.
  - *Hamate*: Distal to triquetral, and the hamulus (hook) of the hamate should be very gently palpated by pressure with the thumb. The ulnar nerve lies superficially, and excessive pressure may produce an unpleasant sensation.
  - *Capitate*: Prominent immediately proximal to the third metacarpal when the wrist is flexed.
  - *Trapezoid*: Proximal to the base of the second metacarpal of the index finger.
  - *Trapezium*: This lies at the base of the thumb next to the base of the first metacarpal. Like the scaphoid, it may also be grasped between palpating thumb and index finger, but differentiate between the scaphoid and trapezium.

(a) Palmar aspect

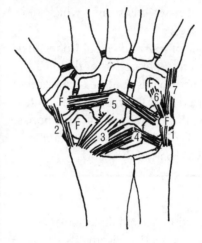

(b) Dorsal aspect

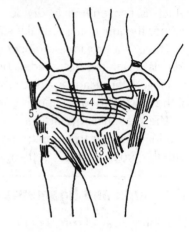

1 Ulnar collateral
2 Radial collateral
3 Palmar radiocarpal
4 Palmar ulnocarpal
5 Palmar intercarpal (deltoid/'V')
6 Pisohamate   } Connected to flexor
7 Pisometacarpal   } carpi ulnaris tendon
F Boundaries of the flexor retinaculum

1 Ulnar collateral
2 Radial collateral
3 Dorsal radiocarpal
4 Dorsal ulnocarpal
5 Pisometacarpal (connected to flexor carpi ulnaris tendon)

(c) Distal radio-ulnar joint (synovial: pivot (selected features))

(Pronation/supination)

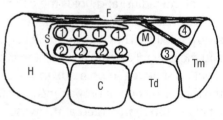

1 Triangular articular disc
2 Recessus sacciformis (synovial membrane pouch)
3 Meniscus
4 Interosseous membrane

(d) The flexor retinaculum and carpal tunnel (transverse section of wrist)

F   Flexor retinaculum
H   Hamate
C   Capitate
Td Trapezoid
Tm Trapezium
M   Median nerve
S   Synovial sheath

Tendons
1 Flexor digitorum superficialis
2 Flexor digitorum profundus
3 Flexor pollicis longus
4 Flexor carpi radialis

Figure 12.2 The ligaments and related structures of the wrist

The function of ligaments in most joints is to limit excessive movement, and in doing so to provide stability. In the case of the wrist, they are particularly important in locking the individual carpal bones together when the hand is in functionally important positions. For example, the palmar ligaments are thicker and stronger than the dorsal ligaments, because the extended wrist is more frequently loaded during hand movements.

The sheer number of articular surfaces present in the wrist (as in the foot) results in an almost bewildering array of ligaments. Taleisnik (1985) has devised a classification of intrinsic and extrinsic ligaments, with the extrinsic ligaments defined as those that link the carpal bones to the radius, ulna or metacarpals. The intrinsic ligaments attach the carpal bones to each other.

Of the palmar extrinsic ligaments, the ulnolunate and radiolunate ligaments (part of the palmar ulnocarpal and radiocarpal complex) form a functional V-shape for strength and support. This arrangement also occurs in the intrinsic ligaments, where the stronger palmar group has a V-shaped ligament that links the capitate to the scaphoid and triquetral. This is the palmar intercarpal ligament, which Taleisnik (1985) has called the deltoid or 'V' ligament (Figure 12.2(a)). Thus, there is actually a double-V system, which is particularly important during abduction and adduction movements of the wrist (pp. 130–131).

Study task

- Refer to Figure 12.2(a), and emphasise the shading of the double-V system of intrinsic and extrinsic ligaments on the palmar surface of the wrist.

Muscles and movement

The wrist is traversed by no less than ten muscle tendons, but practically none of these attaches to the carpal bones. Thus, in the case of the wrist, the protective stability that muscles usually offer joints is reduced. The wrist muscle tendons are also vulnerable, since they may move as much as 3–4 cm through protective tendon sheaths, which are prone to injury if the wrist is repetitively traumatised, for example in sport or repetitive work tasks.

Movement occurs in two planes: flexion/extension and abduction/adduction, and combinations of these. Circumduction is also possible at the wrist. Rotation is not actively possible, but is achieved, in effect, by the pronation/supination action of the radioulnar joints, with contributions from the shoulder if further rotation is required.

The wrist flexor and extensor muscles also control abduction and adduction (radial and ulnar deviation, respectively), easily remembered since the 'radialis' flexor and extensor muscles produce radial deviation, while the 'ulnaris' group produces ulnar deviation.

Flexion

The articular surface of the distal radius has a slight palmar tilt and flexion tends to exceed extension by about 10°. Approximately 60% of the movement takes place at the mid-carpal joint, but variations may occur between individuals (Sarrafian *et al.*, 1977).

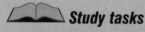

Study tasks

- Highlight the names of the muscles producing the movement.
- Muscle check.
- Palpate the movement in your own wrist.
- Consider which tissues limit flexion.

0–85° (mainly at midcarpal joint)

Movement produced by:
- flexor carpi radialis
- flexor carpi ulnaris
- palmaris longus

assisted by:
- flexor digitorum superficialis and profundus
- flexor pollicis longus
- abductor pollicis longus

Figure 12.3 Flexion of the wrist

Extension

Extension is usually more restricted than flexion, with the greatest contribution taking place at the radiocarpal joint.

Study tasks

- Highlight the names of the muscles producing the movement.
- Muscle check.
- Palpate extension in your own wrist comparing the movement with flexion.
- Consider which tissues limit extension, and whether in reality the range of extension matches flexion.

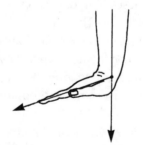

0–85° (mainly at the radiocarpal joint)

Movement produced by:
- extensor carpi radialis longus
- extensor carpi radialis brevis
- extensor carpi ulnaris

assisted by:
- extensor digitorum
- extensor digiti minimi
- extensor indicis
- extensor pollicis longus

Figure 12.4 Extension of the wrist

Abduction (radial deviation)

During this movement, the scaphoid moves closer to the radius, and complex rotatory movements take place in the carpal bones. The movements are both aided, but also ultimately restrained, by the intrinsic and extrinsic ligaments, and particularly the double-V arrangement (Taleisnik, 1985). Abduction takes place mainly at the mid-carpal joint.

0–15° (mainly at the mid-carpal joint)

Movement produced by:
- flexor carpi radialis
- extensor carpi radialis longus and brevis

assisted by:
- abductor pollicis longus
- extensor pollicis brevis

Figure 12.5 Abduction of the wrist

 Study tasks

- Highlight the names of the muscles producing the movement.
- Muscle check.
- Palpate the movement in your own wrist.
- Consider which tissues limit abduction.

Adduction (ulnar deviation)

During this movement, the scaphoid moves away from the radius, with complex rotatory movements in the carpal bones under ligamentous control. The triquetral shifts closer to the ulna, and in extreme adduction may come into contact with the bone, which is normally prevented by the articular disc. Adduction takes place mainly at the radiocarpal joint.

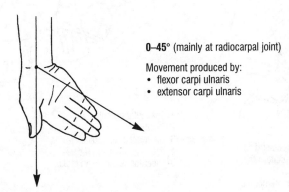

0–45° (mainly at radiocarpal joint)

Movement produced by:
- flexor carpi ulnaris
- extensor carpi ulnaris

Figure 12.6 Adduction of the wrist

 Study tasks

- Highlight the names of the muscles producing the movement.
- Muscle check.
- Palpate adduction in your own wrist, comparing the movement with abduction.
- Consider the tissues that limit adduction, and consider why adduction is so much greater than abduction.
- Consider the practical applications and functional importance of abduction/adduction movements.

Study tasks

- Perform all movements with your own wrist.
- Muscle check.
- Consider which position of the wrist is likely to be the close-packed position and why.

Study tasks

- Highlight the details shown.
- Revise the study tasks relating to pronation/supination on p. 124
- Consider which tissues limit pronation and supination.

Circumduction

This is a combined movement of flexion/adduction/extension/abduction, in that order, or the reverse.

Pronation/supination at the distal radio-ulnar joint

The distal radius possesses a notch (the ulnar notch) which allows a rotatory movement around the head of the ulna. The two bones are connected by a biconcave triangular fibrocartilage disc, which binds the distal radius to the root of the ulnar styloid process. The proximal surface of the disc articulates with the head of the ulna, while the distal surface articulates with the medial part of the lunate, and with the triquetral if the hand is adducted. The synovial membrane that lines the capsule of the joint projects upwards to form a pouch (recessus sacciformis). This lies in front of the lower section of the interosseous membrane that links the shafts of the radius and ulna between the two radio-ulnar joints. As between the tibia and fibula in the leg, the interosseous membrane provides stability and is an important point of attachment for muscles. In the forearm, it is taut in the midway position but relaxed in the positions of pronation and supination.

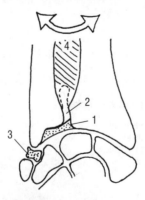

(a) Distal radio-ulnar joint (synovial: pivot)

(Pronation/supination)

1 Triangular articular disc
2 Recessus sacciformis
3 Meniscus
4 Interosseous membrane

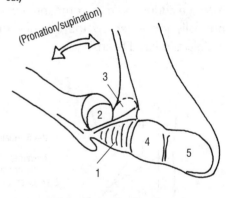

(b) Distal radio-ulnar joint (oblique view – opened out)

(Pronation/supination)

1 Triangular articular disc
2 Head of ulnar
3 Ulnar notch on radius
4 Facet for lunate on radius
5 Facet for scaphoid on radius

Figure 12.7 Pronation/supination at the distal radio-ulnar joint

Stability of the wrist

The fact that so many muscles cross the wrist, yet do not attach to the carpal bones, means that stability of the wrist must be achieved in other ways, facilitated by the following factors:

- The individual carpal bones are multifaceted, giving tight fit in a number of wrist positions.
- The double-V system of ligaments formed by the radiolunate, ulnolunate and palmar intrinsic ligaments (Figure 12.2(a–b)) gives stability, particularly in abduction, adduction and extension.
- The muscles exert a compressive action drawing the carpal bones tightly together.
- If the hand/wrist complex is held steady, the extensor muscles oppose the flexor muscles and therefore stabilise the position. According to Steindler (1955) the following muscles act as fixators: (i) the extensor digitorum and extensor indicis oppose the flexor carpi radialis and flexor pollicis longus; (ii) the extensor carpi ulnaris opposes the extensor pollicis brevis; (iii) the abductor pollicis longus and extensor carpi radialis longus oppose the flexor carpi ulnaris and flexor pollicis longus.
- In addition, Kauer (1980) has suggested that the lateral and medial muscles, abductor pollicis longus and extensor pollicis brevis (radial side) and the extensor carpi ulnaris (ulnar side) provide important collateral support.

The position of the wrist in hand movement

The main function of the wrist is to allow the hand to be placed in a position of optimum efficiency, especially for the precise control of finger and thumb movements.

On completion of the Study Task you should have discovered that finger flexion is more powerful with the wrist held in extension, since this allows the flexor tendons optimal excursion. It is also the close-packed position for the wrist (Williams, 1995). For the same reason, flexion of the wrist hinders the range of thumb movement.

Volz *et al.* (1980) have suggested that maximum grip strength occurs with the wrist in 20° extension, and least with the wrist in 40° flexion. Hazelton *et al.* (1975) found that ulnar deviation, as well as slight extension, produced maximal grip strength. According to Linscheid (1986) this position allows the greatest contact for the radiocarpal joint, which is the main axis for wrist extension and ulnar deviation. Thus, the optimal functional position for the hand

 Study task

- Consider the grip strength of the hand in the various wrist positions of flexion and extension combined with abduction and adduction.

(for most activities) would appear to require the wrist to be held in approximately 20° of extension.

Clinical Note

(a) *This strength may become a point of vulnerability if the wrist is held in extension for too long when performing repetitive tasks such as keyboard work with the finger flexor muscles. It is the supporting extensor muscles which are vulnerable to fatigue and strain, leading to repetitive strain injury (RSI).*

(b) *Falls on an outstretched hand will tend to fracture the scaphoid and distal radius when the wrist is protectively locked in extension. Since the blood supply to the scaphoid tends to favour the distal part of the bone, there is a real danger of avascular necrosis and non-union to the middle and proximal parts of a fractured scaphoid.*

The lunate is vulnerable to dislocation or subluxation, owing to its shape and pivotal position. Posterior displacement is not uncommon; but in a fall on an outstretched hand the weaker dorsal ligaments may be torn and the lunate rotated through 90° by compression from the radius and capitate. It tends to hinge on the palmar ligaments and shift anteriorly, where it may compress the median nerve.

In spite of the fact that extension of the wrist provides the optimal position for many hand functions, Volz *et al.* (1980) found that in terms of pure wrist motion the most powerful muscle is the flexor carpi ulnaris. Wrist flexion also allows considerable range of movement, and when combined with supination/pronation is important in gestures and sporting skills. It allows spin to be imparted to both bat and ball, and enables the strings of a cello to be caressed with a bow.

In terms of shock absorption, the wrist is designed to withstand the kinds of compressive force that frequently occur in falls to the outstretched hand (not always injurious) or in punching movements. The transmission of force passes through the stable second and third metacarpal bones to the capitate, where it dissipates to the scaphoid and lunate, and thence to the radio-ulnar joint, the tri-angular fibrocartilaginous disc being of considerable importance in absorbing these forces (Palmer and Werner, 1984).

Chapter 13

The hand

Introduction

The hand is a remarkably versatile feature of human anatomy. Many of the achievements of human civilisation reflect the skill and versatility of the hand, and the main purpose and function of the other joints in the upper limb are to allow the hand optimum movement and orientation.

The terminology used here will be defined at the outset. The term 'digit' is often used to describe the fingers, but the thumb should also be included. The term 'ray' may be usefully employed to describe the functional chain consisting of the relevant metacarpal, phalanges and intervening joints. From radial (lateral) to ulnar (medial) they are numbered and designated as follows:

I   thumb
II   index finger
III   middle finger
IV   ring finger
V   little finger

The bones and joints

(a) Bones and joints of the hand

① – ⑤ Anatomical designation of digits

1 Metacarpals
2 Proximal phalanges
3 Middle phalanges
4 Distal phalanges

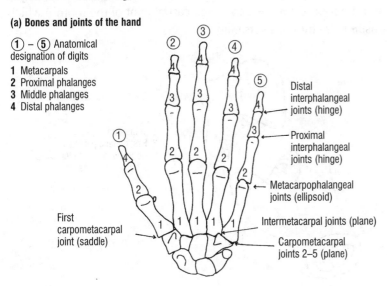

First carpometacarpal joint (saddle)

Distal interphalangeal joints (hinge)

Proximal interphalangeal joints (hinge)

Metacarpophalangeal joints (ellipsoid)

Intermetacarpal joints (plane)

Carpometacarpal joints 2–5 (plane)

(b) The arches of the hand

(i)   a fairly rigid transverse carpal arch
(ii)   a mobile transverse metacarpal arch (at the knuckles)
(iii) a rigid longitudinal metacarpal arch formed by the shafts of the metacarpal bones
(iv)   a mobile interphalangeal arch formed by the fingers

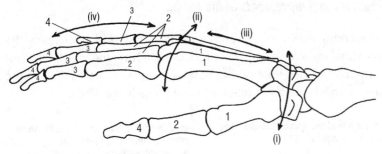

1 Metacarpals
2 Proximal phalanges (sing. phalanx)
3 Middle phalanges
4 Distal phalanges

Figure 13.1 The bones, joints and arches of the hand

The significance of the thumb

The thumb is 'worth half the hand'. The truth underlying this simple assertion may be tested by trying to perform a number of everyday tasks without using the thumb. Efficiency is greatly impaired.

The thumb ray consists of the first metacarpal and proximal and distal phalanx (there is no middle phalanx in the thumb) (Figure 13.1(a)). However, it is the saddle joint between the first metacarpal and the trapezium (the carpometacarpal joint of the thumb) which ensures the thumb's versatility.

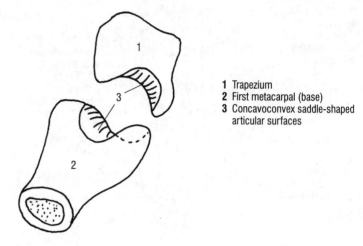

1 Trapezium
2 First metacarpal (base)
3 Concavoconvex saddle-shaped
  articular surfaces

Figure 13.2 The bones of the carpometacarpal joint of the thumb (oblique view – opened out)

Muscles and movements of the thumb

The configuration of the arches of the hand, and particularly the proximal transverse carpal arch (Figure 13.1(b)), means that the first carpometacarpal joint of the thumb is medially rotated, and so the axes of movement are different from those in the fingers.

(a) Flexion of the thumb (includes some medial rotation) (0–50°)

Movement produced by:
• flexor pollicis brevis
• opponens pollicis
• flexor pollicis longus

(b) Extension of the thumb (includes some lateral rotation (0–50°) from the anatomical position)

Movement produced by:
• extensor pollicis longus
• extensor pollicis brevis
• abductor pollicis longus

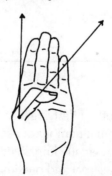

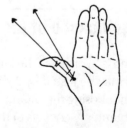

(c) Abduction of the thumb (0–50°)

Movement produced by:
- abductor pollicis longus
- abductor pollicis brevis

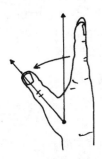

(d) Adduction of the thumb (50–0°)

Movement produced by:
- adductor pollicis

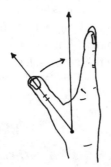

(e) Opposition of the thumb

Movement produced by:
- opponens pollicis
- flexor pollicis brevis

(f) Circumduction of the thumb

Movement produced by: consecutive action of
- extensors
- abductors
- flexors
- adductors

(g) The muscles of the thenar and hypothenar eminence

The thenar eminence:
1 Opponens pollicis
2 Abductor pollicis brevis
3 Flexor pollicis brevis

The hypothenar eminence:
4 Opponens digit minimi
5 Flexor digiti minimi brevis
6 Abductor digiti minimi

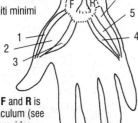
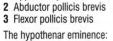
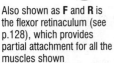

Also shown as **F** and **R** is the flexor retinaculum (see p.128), which provides partial attachment for all the muscles shown

(h) Cupping of the hand

Figure 13.3 The muscles of the thenar and hypothenar eminence

According to Zancolli *et al.* (1987), the most stable position for the first carpometacarpal joint of the thumb is that of opposition, combined with pronation of the hand and flexion and adduction of the thumb. This position tightens the ligaments – and especially the strong palmar ligament on the ulnar side – that link the trapezium with the base of the first metacarpal, and maximises the stabilising role of the thenar muscles.

Clinical Note

The thumb is less stable in extension, and the important palmar ligament of the thumb may be damaged if it is violently extended. Skiers may suffer this injury if the thumb is entangled in a ski pole during a fall ('skier's thumb').

In addition to the intrinsic muscles of the thenar eminence, it should also be noted that some of the muscles that move the thumb are extrinsic to the hand and originate in the forearm.

The 'anatomical snuffbox'

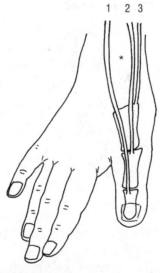

* The anatomical snuffbox
Tendons
1 Extensor pollicis longus
2 Extensor pollicis brevis
3 Abductor pollicis longus

Figure 13.4 The 'anatomical snuffbox'

Study tasks

- Shade the tendons shown in Figure 13.4 which form the 'anatomical snuffbox'.
- Palpate the tendons with your thumb and fingers in full extension. Note that the two lateral tendons are enclosed by a single synovial sheath and are difficult to palpate individually. The hollow formed by the anatomical snuff box allows direct palpation of the scaphoid, which is liable to fracture in falls on an outstretched hand.
- Also identify and palpate the distal end of the radius and its styloid process. Note the gap between the radius and the carpal bones when the wrist is relaxed, and that it is obliterated if the thumb is extended. Identify the tendons that obliterate this gap. Repetitive strain of these tendons may give rise to the inflammatory condition known as 'de Quervain's tenosynovitis'.
- Locate the trapezium and try to palpate the joint line between the trapezium and base of the first metacarpal. This is the first carpometacarpal joint of the thumb, and a common site of degenerative change in the elderly.

The metacarpophalangeal joint of the thumb

The head of the first metacarpal articulates with the base of the proximal phalanx in a semicondylar joint which is sometimes classified as ellipsoid (Williams, 1995). The movements are essentially flexion/extension and abduction/adduction, but a degree of passive rotation can be demonstrated.

In common with the other metacarpophalangeal joints, there is a fibrocartilaginous volar plate on the palmar surface (see under 'The metacarpophalangeal joints' below). There are also two sesamoid bones present in the capsule and associated with the tendons of the flexor pollicis brevis and adductor pollicis muscles.

The interphalangeal joint of the thumb

Similar in structure to the interphalangeal joints of the fingers, the interphalangeal joint of the thumb allows flexion and extension, with a degree of passive rotation, as pronation of the wrist with opposition of the thumb is achieved. The movement actually involves the entire thumb ray at the first carpometacarpal (saddle) joint, at the metacarpophalangeal joint and then finally at the interphalangeal joint.

The finger rays (II–V)

The carpometacarpal joints

Note that the greatest mobility is found in the thumb and little finger carpometacarpal joints. Least mobility is found in the index and middle finger carpometacarpal joints. The latter form the stable part of the hand, demonstrable in punching with a closed fist.

The metacarpophalangeal joints

Williams (1995) classifies these as ellipsoid joints but acknowledges that on their palmar surface the metacarpal heads are partially divided and virtually bicondylar. Elsewhere they have been called 'multiaxial condyloid' (Hertling and Kessler, 1990). All agree that flexion/extension and abduction/adduction are possible, but the latter term allows for a degree of rotation, which probably reflects a relatively loose capsule and shallow joint surfaces.

The most notable anatomical feature of these joints is the presence of a complicated structure forming the palmar ligament,

Study tasks

- Examine passive rotation of the thumb in both flexion and extension. Consider why it is functionally useful to have more rotation in flexion than in extension.
- Compare variations in the passive range of movement of the thumb joints of your colleagues; note that considerable variation is normal.

Study tasks

- With reference to Figure 13.1(a), grasp each metacarpal bone commencing with the thumb. This has already been described as mobile. Compare the mobility of each metacarpal in turn, noting the differences.
- Locate and identify each carpal bone associated with each finger ray.
- Make a loose fist, and then a tightly closed fist. Explain the difference in appearance.

which is often referred to as the 'volar plate'. It consists of a plate or tab of fibrocartilage that is inserted into the (volar) base of each proximal phalanx, including the thumb. Its surface is grooved for the flexor tendons, whose synovial sheaths are attached to the sides of the grooves. The palmar ligament then becomes thinner and membranous as it attaches to the metacarpal, and acts as a 'check' ligament opposing hyperextension (Figure 13.5(a)).

Support at the sides is provided by cord-like collateral ligaments. On the palmar surface there is a groove for the flexor tendons, which are enclosed by a fibrous sheath; in flexion the membranous part of the volar plate folds rather like a concertina. On the dorsal surface, ligaments apart from the capsule are largely replaced by the extensor tendons, which are separated from the capsule by small bursae.

The second, third, fourth and fifth metacarpophalangeal joints are linked by deep transverse metacarpal ligaments (Figure 13.5(b–c)) and the superficial transverse metacarpal ligaments that lie above.

(a) In flexion (0–90°) (lateral view)

(b) In extension (oblique view)

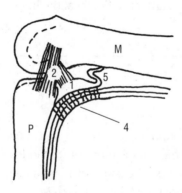
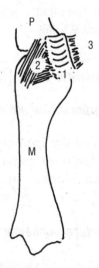

(c) In extension (palmar view)

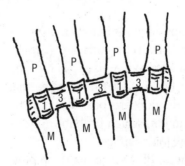

1 Palmar ligament (volar plate)
2 Collateral ligaments (taut in flexion, more relaxed in extension)
3 Deep transverse metacarpal ligaments
4 Flexor sheath enclosing flexor tendons
5 'Check' ligaments
M Metacarpal
P Proximal phalanx

Figure 13.5 The metacarpophalangeal joints (synovial: ellipsoid, bicondylar)

Note: Movements at the metacarpophalangeal joints are flexion/extension, abduction/adduction and circumduction; rotation occurs as the fingers close around an object, but cannot be activated independently. These movements and the muscles producing the movements will be discussed under 'Movements of the hand', below.

The interphalangeal joints

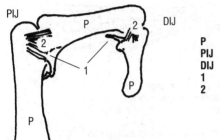

P	Phalanx
PIJ	Proximal interphalangeal joint
DIJ	Distal interphalangeal joint
1	Palmar ligament
2	Collateral ligament

Study tasks

- Examine the amount of flexion/extension available in your own interphalangeal joints.
- Highlight the features shown on the diagrams.
- Muscle check.

Figure 13.6 The interphalangeal joints (lateral view) (synovial: hinge)

The proximal interphalangeal joints have a similar ligamentous arrangement to the metacarpophalangeal joints, except that the flexor sheath enclosing the flexor tendons inserts on to the volar plate and on to both proximal and distal phalanges. This gives increased stability in extension of the fingers. In the case of the metacarpophalangeal joints, the flexor sheath does not attach to the metacarpal bones.

The distal interphalangeal joints have a similar structure except that the volar plate does not possess check ligaments proximally, allowing greater extension (Sandzen, 1979). Flexion is greatest at the proximal interphalangeal joints, while extension is greater at the distal joints. Apart from the joint capsules, dorsal ligaments are absent and are largely replaced by the extensor tendon mechanism described below.

The dorsal digital expansion (extensor expansion/mechanism)

This is a broad triangular aponeurosis consisting of the main extensor tendons that pass over the dorsal surface of the finger rays, and the lateral attachments where the tendons of the interosseous and lumbrical muscles join. The extensor tendons are therefore essentially extrasynovial, and exert their main force at the metacarpophalangeal joints. Of particular interest is the way in which the lateral tendons of the intrinsic lumbrical and interossei muscles are able to flex the metacarpophalangeal joints of the fingers, yet extend the interphalangeal joints. They achieve this by inserting distal to the metacarpophalangeal joints (producing flexion), and dorsal to the axis of movement of the interphalangeal joints (producing extension).

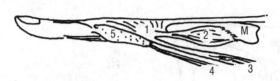

1 Dorsal digital expansion
2 Palmar interosseous muscle (interossei)
3 Lumbrical muscle attached to flexor
    digitorum profundus tendon
4 Tendon of flexor digitorum superficialis
5 Fibrous flexor sheath
M Metacarpal

Figure 13.7 The dorsal digital expansion (lateral view of fingers)

Movements of the hand

Although it is important to understand the structure and function of each type of joint in the hand, it is also important to realise that individual joint movements are usually part of functional movements of the hand as a whole. Figure 13.8 shows the general functional movements of the hand, but even these offer little insight into the ingenuity of use to which the prehensile human hand may be put (see Figure 13.9).

Flexion takes place at the metacarpophalangeal and interphalangeal joints and involves slight conjunct lateral rotation of the second–fifth fingers, but with medial rotation of the thumb. These

---

(a) Flexion of the hand

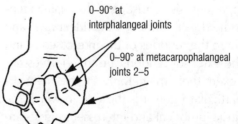

0–90° at
interphalangeal joints

0–90° at metacarpophalangeal
joints 2–5

Movement produced by:
• flexor digitorum superficialis
• flexor digitorum profundus
• flexor pollicis longus
• flexor digiti minimi brevis

assisted by:
• lumbricals
• interossei
See also 'Flexion of the thumb'
(p.138)

Study tasks

• Highlight the details shown in Figure 13.7 and ensure that you understand how the lumbrical and interossei muscles work.
• With reference to delicate finger movements, consider the significance of metacarpophalangeal flexion combined with interphalangeal joint extension.
• Consider which muscles activate the thumb and little finger during these movements.

(b) Extension of the hand (0–40°)

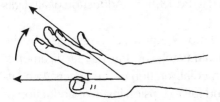

Movement produced by:
- extensor digitorum
- extensor indicis
- extensor digiti minimi
- extensor pollicis longus and brevis

(c) Abduction of the hand (accompanies extension)

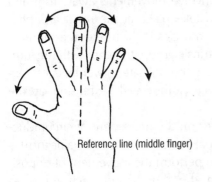

Reference line (middle finger)

Movement produced by:
- dorsal interossei
- abductor digiti minimi
- abductor pollicis brevis

assited by:
- extensor digitorum
- extensor indicis
- extensor digiti minimi

(d) Adduction of the hand (accompanies flexion)

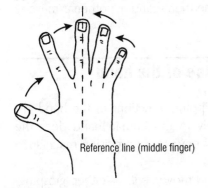

Reference line (middle finger)

Movement produced by:
- palmar interossei
- adductor pollicis

assisted by:
- flexor digitorum superficialis
- flexor digitorum profundus

(e) Opposition of the little finger

Movement produced by:
- opponens digiti minimi

Figure 13.8 Movements of the hand

Study tasks

- Highlight the details shown in Figure 13.8.
- Consider the role of the individual joints in each movement shown.
- Consider which tissues limit each movement shown.
- Muscle check.

conjunct rotation movements are extremely useful in the pincer-like movement of opposition (Figure 13.8(e)). Adduction of all digits tends to accompany flexion.

Extension is a more limited movement which is restricted by the flexor muscle tendons, and ultimately by the palmar ligaments. Conjunct rotation produces medial rotation of the fingers with lateral rotation of the thumb; and full extension involves abduction of the hand due to the line of pull exerted by the extensor muscles.

Abduction and adduction involve movements away from and towards the middle finger and take place at the metacarpophalangeal joints. The interossei muscles, which are the primary movers, also form part of an intrinsic group (together with the lumbricals) responsible for producing the delicate movements of flexion at the metacarpophalangeal joints, with extension at the interphalangeal joints.

Each digit is also able to perform the movement of circumduction at the metacarpophalangeal joint, but the real value of this freedom is to allow the hand to perform the movement of opposition (Figure 13.8(e)) between the thumb and each finger as required. This is due not only to the structure of the individual joints, but also to the action of the thenar and hypothenar muscles combined (Figure 13.3(g)).

The functional significance of the hand

The prehensile (grasping and gripping) functions of the hand are seen in other primates, but only in the human hand does the degree of refinement in terms of power and precision reach such impressive levels.

Kapandji (1982) classifies hand movements as either 'grasping' or 'non-grasping' and describes four types of grasping actions.

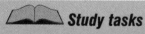

Study tasks

- Think of as many 'non-grasping' actions of the hand as you can.
- Adopt the routine described (p. 147) (by Napier) as you grip an object such as a hammer or a tennis racket.
- Analyse the movements and muscles used at each stage in the process.
- An expert will usually make refinements in order to improve efficiency and avoid injury. Discover as many faults/injuries as you can that are associated with incorrect grip tendencies in sports or work-related use of hand-held equipment.

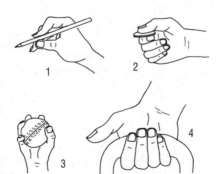

1 The chuck grip which uses at least two fingers and the thumb and operates like a drill chuck
2 The pliers grip between thumb and index finger
3 The ring grip which surrounds an object
4 The hook grip which is less precise, and is the only grip that does not rely heavily on the thumb

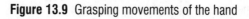

Figure 13.9 Grasping movements of the hand

Napier (1956) distinguishes between a power grip and a precision grip, suggesting the following sequence:

1. Opening the hand.
2. Positioning the fingers (anticipation).
3. Approaching and closing around an object.
4. Adopting a strong and static grip.

Precision handling involves the same sequence but usually involves holding the fingers and thumb in a position of opposition, using varying degrees of contact between the pads or pulp area of the fingers, or even using the tips and nails. Hence the inevitable loss of precision among people who bite their nails.

The role of the wrist in hand movements

It will be evident when adopting various gripping positions with the hand that the position of the wrist is vital. This is because the wrist provides a stable platform for the hand, and its position also controls the length of the forearm muscles that operate the fingers.

For example, flexion of the fingers requires that the wrist be held in extension. The opposite is true if strong finger or thumb extension is required.

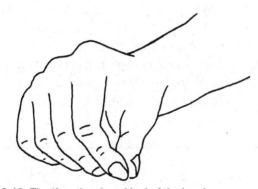

Figure 13.10 The 'functional position' of the hand

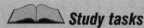 **Study tasks**

- Attempt to pick up a pin, and make some light downstrokes with a paintbrush.
- Analyse the movements and muscles used at each stage in the process.

 Study tasks

- Try to grip an object strongly with the wrist in varying degrees of flexion/extension and in ulnar and radial deviation. Consider which position feels most effective, and try to explain why.
- Perform a task that involves extending the fingers (try brushing an insect away with the dorsum of the hand). Consider which wrist position is most effective, and try to explain why.

Your answers may help to explain why the so-called 'functional position of the hand' involves approximately 20° wrist extension and 10° ulnar deviation, with fingers and thumb in a position of anticipatory opposition.

Clinical Note

The position shown in Figure 13.10 is often regarded as the optimum position for efficient hand function, and is usually the position of choice if either hand or wrist is immobilised following injury.

Chapter 14

The hip

Introduction

At first glance the upper and lower limbs have a similar basic design, and this reflects the fact that humans are continuing to evolve from a quadrupedal towards a bipedal stance, in which the upper limb generally performs specialised non-weightbearing tasks. The lower limb, however, remains primarily as a weightbearing and locomotive structure which supports the upright body.

Function tends to influence structure, and in the case of the hip there is evidence that the structure is still evolving and that it has not yet fully caught up with the change in function from a quadrupedal to a bipedal stance. Nevertheless, the hip is a large and stable joint, and in health is able to absorb impressively large forces. It is probably due to this that any derangement in functional anatomy (for example in terms of altered weightbearing) can accelerate the process of degenerative change that is unfortunately common in the hip joint.

The bones

The contact surfaces of the hip joint consist of a socket formed by the acetabulum (socket) of the pelvis and the rounded head of the femur, which forms approximately two-thirds of a sphere.

If a skeletal specimen or model is examined, it is noticeable that the deepest fit is obtained with the shaft of the femur in a degree of abduction, flexion and lateral rotation, compared with the so-called anatomical position.

The anatomical position is that assumed in a normal bipedal stance, and observation of this position reveals that the weightbearing portion of the femoral head is confined to an area on the posterior and superior surface, and that a surprisingly large area on the anterior surface does not appear to be weightbearing at all. Thus, in the standing position, the acetabulum faces obliquely anteriorly, laterally and inferiorly.

(a) Anterior view

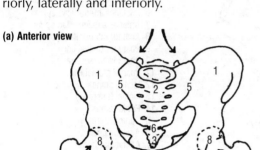

(b) Posterior view

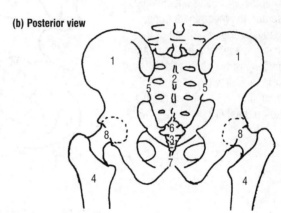

Weightbearing forces

Bones	Joints
1 Innominate	**5** Sacro-iliac joint (synovial)
2 Sacrum	**6** Sacro-coccygeal joint (2° cartilaginous)
3 Coccyx	**7** Pubic symphysis (2° cartilaginous)
4 Femur	**8** Coxal (hip) (ball and socket)

(c) The upper end of the femur

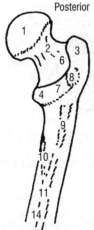

Anterior

Posterior

1 Head
2 Neck
3 Greater trochanter
4 Lesser trochanter
5 Intertrochanteric line
6 Trochanteric fossa
7 Intertrochanteric crest
8 Quadrate tubercle
9 Gluteal tuberosity
10 Spiral line
11 Linea aspera
12 Angle of inclination (125°)
13 Angle of torsion (varies 10–30°)
14 Shaft

Superior

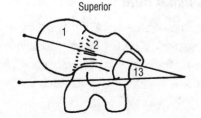

Study tasks

- Highlight the bone features shown, and identify the features on bone specimens where possible.
- Try to obtain matching specimens of a femur and innominate bone, and carefully align the head of the femur in the acetabulum. Compare the position of the shaft of the femur in the following situations: (i) the anatomical position; (ii) the position of best fit.

(d) The right innominate bone (lateral view)

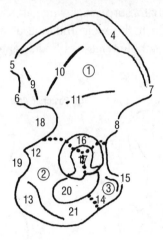

① ilium
② ischium
③ pubis

Limits of iliac, pubic and ischial bones

4 Iliac crest
5 Posterior superior iliac spine
6 Posterior inferior iliac spine
7 Anterior superior iliac spine
8 Anterior inferior iliac spine
9 Posterior gluteal line
10 Anterior gluteal line
11 Inferior gluteal line
12 Ischial spine
13 Ischial tuberosity
14 Pubic ramus
15 Pubic tubercle
16 Acetabulum
17 Acetabular notch
18 Greater sciatic notch
19 Lesser sciatic notch
20 Obturator foramen
21 Ischial ramus

(e) The trabecular structure (trabeculae)

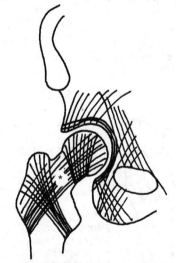

* Weak zone

Figure 14.1 The hip and pelvis: main bone and joint features

(b) The internal trabecular structure of cancellous bone in the femur shows a zone of relative weakness in the neck (Figure 14.1(e)), which is a common site of fractures in the elderly.

(c) Differences in the angle of inclination (see below) are one reason for leg length discrepancies. Increases and decreases in the angle of torsion (which in the hip is anteverted) will cause the knee to rotate medially (increased anteversion) or laterally (decreased anteversion) (Figure 14.2 (c), (d)). Bear in mind that changes in the angle of torsion along the shaft of the femur will tend to be expressed at the distal end of the bone, at the knee. The position of the femoral head in the acetabulum will remain relatively constant.

(a) Coxa valga   **(b) Coxa vara**

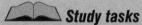

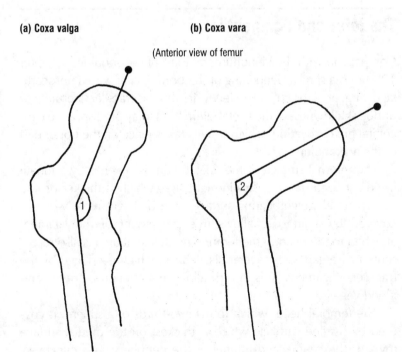

Study tasks

- With reference to a bone specimen and Figure 14.2(c), consider the effects of an increase in the angle of inclination of the femur (the condition known as 'coxa valga') and a decrease in that angle ('coxa vara').

Figure 14.2 Variations in the angles of inclination (a, b) and torsion (c, d)

Study tasks

- Repeat the Study Task on p. 151 for the angle of torsion.
- Refer back to the angle of torsion in the humerus (Figure 10.8(b), p. 109) and explain why the femur is 'anteverted' but the humerus 'retroverted'.

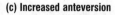

(c) Increased anteversion

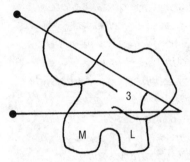

(d) Decreased anteversion

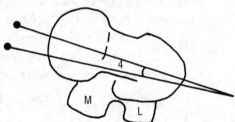

1   More than *c.* 125°
2   Less than *c.* 125°
3   More than *c.* 30° anteversion
4   Less than *c.* 10° anteversion
M   Medial femoral condyle (knee)
L   Lateral femoral condyle (knee)

Figure 14.2 (continued)

The joint and ligaments

Comparison with the glenohumeral joint of the shoulder (Chapter 10) shows a similar deepening of the bony socket with a fibrocartilage ring or 'labrum'. Deficiency in this 'acetabular labrum', or undue shallowness of the acetabulum itself, may predispose to congenital hip dislocation, which usually takes place on the upper part of the acetabulum.

The internal surface of the acetabulum is covered by a lunate (moon-shaped) layer of hyaline cartilage, which is thicker on the superolateral weightbearing surface. The deficient area, which is covered by a fat pad, allows free movement of the ligament attached to the head of the femur. The deficient area is also represented by a gap in the acetabular labrum which is bridged by the transverse ligament. This in turn allows penetration by nerve and blood vessels.

The femoral head, which forms two-thirds of a sphere, is covered by hyaline cartilage which is thickest on the medial surface towards the centre, and thinner at the periphery. This effectively deepens the socket and promotes stability.

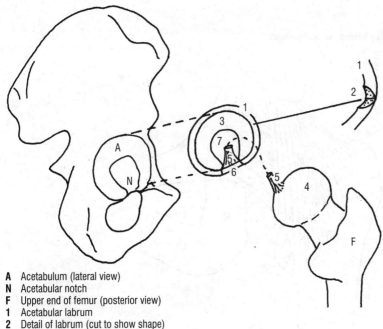

A Acetabulum (lateral view)
N Acetabular notch
F Upper end of femur (posterior view)
1 Acetabular labrum
2 Detail of labrum (cut to show shape)
3 Lunate surface of articular hyaline cartilage
4 Articular hyaline cartilage
5 Ligament of the head of the femur
6 Transverse ligament of the acetabulum

Figure 14.3 The hip joint: selected aspects (opened out)

The ligament of the head of the femur attaches to the 'fovea' which is a small depression in the head of the femur, deficient in hyaline cartilage.

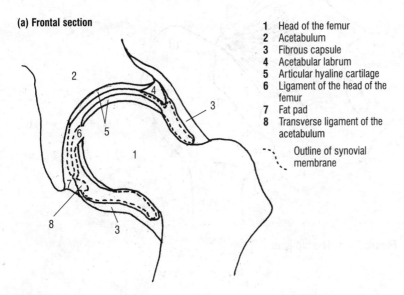

(a) Frontal section

1 Head of the femur
2 Acetabulum
3 Fibrous capsule
4 Acetabular labrum
5 Articular hyaline cartilage
6 Ligament of the head of the femur
7 Fat pad
8 Transverse ligament of the acetabulum

- - - Outline of synovial membrane

📖 ***Study tasks***

- Study Figure 14.4(a) and lightly shade the features shown, in separate colours.
- Study Figure 14.4(b) and (c) and shade the three main ligaments of the hip joint.
- Stand with your hands on your hips (with your thumbs resting on the anterior superior iliac spine of both innominate bones, and fingers pointing backwards resting on the gluteal muscles). Allow your pelvis to rotate posteriorly while allowing your body to sway forwards slightly. In spite of the fact that your muscles are assisting, you are being supported mainly by the very strong iliofemoral ligament.
- Place a colleague in the sidelying position and ask them to hold their lower hip and knee in flexion while you support their upper limb and move their hip into extension. Through your own proprioception feel the end point of the movement. This is caused by the resistance of the iliofemoral ligament, although you are likely to meet some resistance from the hip flexor muscles as well.
- Consider why it is important to ask your colleague to flex the lower limb.
- From the same position take your colleague's hip joint into internal rotation, abduction and extension. Feel the joint tighten up as it reaches the close-packed position. The spiral configuration of the iliofemoral and ischiofemoral ligaments tightens on internal rotation and extension, while the pubofemoral ligament tightens on abduction as well as internal rotation.

(b) Anterior view: ligaments

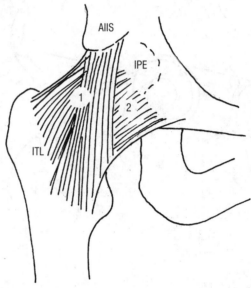

AIIS	Anterior inferior iliac spine
IPE	Iliopectineal eminence
ITL	Intertrochanteric line
IT	Ischial tuberosity
GT	Greater trochanter
ITC	Intertrochanteric crest
1	Iliofemoral ligament
2	Pubofemoral ligament
3	Ischiofemoral ligament

(c) Posterior view: ligaments

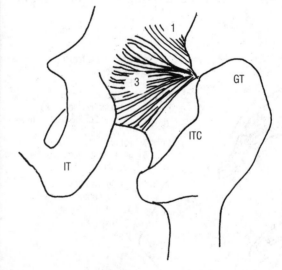

Figure 14.4 The hip joint

The joint capsule is dense and strong, particularly anteriorly, and surrounds the neck of the femur, reinforced by extremely strong ligaments. Several sets of capsular fibres have been identified, some of which run obliquely, some run in an arcuate manner, and some of the deep fibres form a circular arrangement around the neck, known as the 'zona orbicularis'. Other deep fibres lying anteriorly reflect upwards along the neck as longitudinal bands, known as 'retinacula', which accompany blood vessels supplying the joint.

The capsule is lined by synovial membrane, which is extensive, since it surrounds the ligament of the head of the femur, reflects on to the internal surface of the capsule and covers the acetabular labrum and the fat pad in the acetabular fossa. There are three strong ligaments that support the hip joint. These are:

- The iliofemoral ligament (sometimes called the 'Y' ligament due to its shape). It blends with the capsule and runs from the anterior inferior iliac spine of the innominate, to the intertrochanteric line of the femur. It helps to support the body in the standing position, and has a vertical medial band and a lateral oblique band.
- The pubofemoral ligament, which is attached to the iliopectineal eminence of the innominate bone and supports the iliofemoral ligament by passing below and blending with it and the capsule.
- The ischiofemoral ligament, which winds spirally behind the joint and provides posterior support.

It should be noted that all three ligaments tighten in extension, but are relaxed in flexion.

Study tasks

- The ligaments are relaxed in flexion. Test this for yourself from a standing position which allows you to rest on the iliofemoral ligaments. With your pelvis in posterior rotation (flattened abdomen, bottom tucked in), widen your legs into abduction (towards the splits position – take care when attempting this). Reach a point of maximum stretch and then deliberately allow your pelvis to rotate anteriorly, which means that you will be flexing your hips. You will find that you can abduct further because the anterior pelvic rotation allows hip flexion to relax the ligaments. With training, the full splits position can be reached, but is always accompanied by anterior pelvic rotation.

The bursae

These are important in the hip joint, owing to the size and power of the surrounding muscles and tendons. There is a large iliac bursa (sometimes called the 'psoas bursa') which lies under the psoas tendon and may communicate with the synovial membrane anteriorly through a gap between the iliofemoral and pubofemoral ligaments.

Bursae also lie between the tendon of the gluteus maximus and the greater trochanter, over which it passes; and between the greater trochanter and the gluteus medius and minimus muscles. It seems that medieval weavers suffered from this type of bursitis, which they tried to avoid with heavily padded underwear. This was noticeable enough for Shakespeare to pun on the name 'Bottom' for his weaver in *A Midsummer Night's Dream*.

Clinical Note

Blood supply

The role of the head and neck of the femur in weightbearing is so important that the blood supply to these structures assumes vital significance. The main supply comes from the arteries that ascend the neck of the femur, which renders them vulnerable in fractures of the femoral neck. Since they are intracapsular they may also be vulnerable to inflammatory pressures from within the joint. Any interruption of blood flow will prevent bone healing or may predispose to developmental diseases such as osteochondrosis (Perthes' disease of the hip). The head of the femur also receives blood from the acetabular branch of the obturator artery via the acetabular notch and the ligament of the head of the femur; but in as many as 20% of the population this fails to anastomose with the other arteries and it appears to be much less important than the circumflex arteries. The trochanters lie outside the capsule and receive supply from the circumflex branches and the superior and inferior gluteal arteries.

Muscles and movement

The hip is a spheroidal joint which allows freedom of movement through three axes, i.e. flexion/extension, abduction/adduction, internal/external rotation and combinations of these, including circumduction.

Note that due to bipedal stance the hip is already in a degree of extension in the standing position, and only a limited amount of extension is available before the lumbar spine must also be extended in order to increase the effectiveness of the movement. For the same reason, the amount of flexion available seems surprisingly large.

The abductor muscles are used more to support the body during the gait cycle than in pure abduction movements. Adduction will tend to occur from a starting position of abduction, or be combined with flexion or extension, since pure adduction in the anatomical position is not possible.

(a) Flexion of the hip (0–120°)

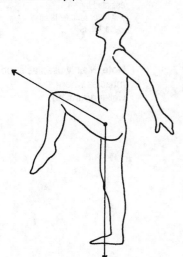

Movement produced by:
- psoas major
- iliacus

assisted by:
- pectineus
- rectus femoris
- sartorius
- adductor muscles
  (in early stages)

 Study tasks

- Highlight the movements shown and the muscles that produce the movements.
- Muscle check.

(b) Extension of the hip (0–40°)

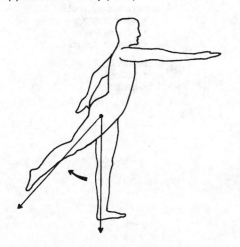

Movement produced by:
- gluteus maximus
- hamstring muscles

Note: The range of extension is greater with an extended knee, since the hamstrings also flex the knee.

(c) Abduction of the hip (0–30°)

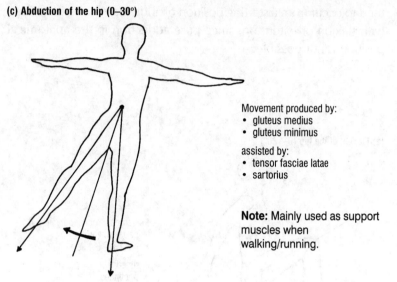

Movement produced by:
- gluteus medius
- gluteus minimus

assisted by:
- tensor fasciae latae
- sartorius

Note: Mainly used as support muscles when walking/running.

(d) Adduction of the hip (shown with flexion) (0–30°)

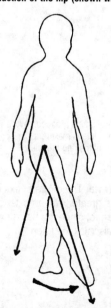

Movement produced by:
- adductor longus
- adductor brevis
- adductor magnus

assisted by:
- pectineus
- gracilis

(e) Medial rotation of the hip (0–30°)

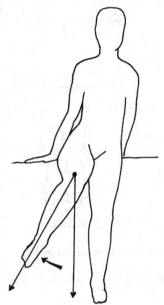

Movement produced by:
- tensor fasciae latae
- gluteus medius and minimus (anterior fibres)

assisted by:
- gracilis
- adductor muscles (limited according to starting position)

 Study tasks

- Consider which of the rotation movements of the hip is more powerful and why.

(f) Lateral rotation of the hip (0–60°)

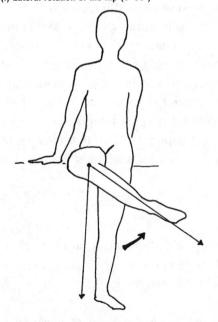

Movement produced by:
- obturator internus
- obturator externus
- gemellus superior
- gemellus inferior
- quadratus femoris

assisted by:
- piriformis
- gluteus maximus
- sartorius

Figure 14.5 Movements of the hip (note: circumduction is not shown)

Clinical Note

The attachments of the abductor muscles allow for an important clinical test of the integrity of the hip, known as the Trendelenburg sign. This is regarded as positive if the unsupported pelvis drops below the level of the supported side when the patient is asked to stand on one leg (Figure 14.6(b)).

Study tasks

- With the aid of a colleague, imitate Trendelenburg positive and negative results.
- Consider the possible reasons for a positive Trendelenburg result.

 Study tasks

- Obtain a matching acetabulum and femur. Place the bones in the anatomical position, and note the anterior direction of both the acetabulum and the neck of the femur.
- Move the femur into flexion and extension in order to obtain pure spin. Consider whether this corresponds to the clinical movements shown above. If not, try to explain why not.
- Keep the acetabulum in the anatomical position and align the femoral shaft in the appropriate degrees of abduction, flexion and lateral rotation which allows alignment of the two mechanical axes. Note that this is a position of best fit and coincides with the quadrupedal position ('on all fours'). It is also a noticeably comfortable position for the hips when sitting.
- Now move the femoral shaft into degrees of extension, abduction and medial rotation. Notice that a position of best fit is obtained once more but this corresponds to the close-packed position. Move the femoral shaft between these two positions and try to decide whether you have pure spin or impure swing.

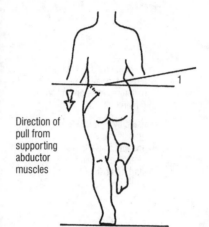

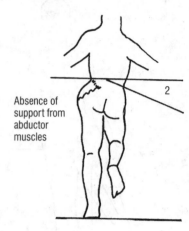

(a) Negative

Direction of pull from supporting abductor muscles

(b) Positive

Absence of support from abductor muscles

1 Negative: Unsupported hip can be lifted higher than weightbearing hip due to normal contraction and strength of abductor muscles.
2 Positive: Unsupported hip cannot be lifted higher than weightbearing hip due to malfunction in abductor mechanism (for whatever reason).

Figure 14.6 The Trendelenburg sign

Further biomechanical considerations

The movements described are those usually recognised for clinical purposes, but in terms of arthrokinematics the situation is more complex. The head of the femur is not truly spherical, and in the anatomical position both the acetabulum and the neck of the femur are directed anteriorly. The dynamic nature of the joint means that the mechanical axis of movement is constantly changing, and although pure spin may be obtained in certain positions of flexion/extension, most movements are pure and impure swings (for explanation of these terms see Chapter 3). Note that in order to obtain pure spin, the neck rather than the shaft must be regarded as the mechanical axis.

The weightbearing hip

The hip joint is a fulcrum upon which the human body pivots. In the normal bipedal standing position the weight passing through each femoral head is approximately half the body's weight minus the weight of the legs.

Standing on one leg results in a shift in the centre of gravity such that the unsupported side would drop if not for the powerful contraction of the gluteal muscles (gluteus medius and minimus) on the supporting side. These attach to the greater trochanter; but the distance from the greater trochanter to the femoral head is only about half the distance of that from the femoral head to the centre of gravity.

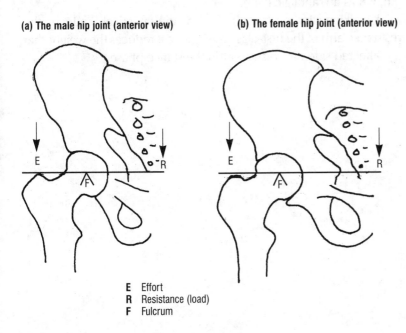

(a) The male hip joint (anterior view)   **(b) The female hip joint (anterior view)**

E   Effort
R   Resistance (load)
F   Fulcrum

Figure 14.7 The hip joint: a first-class lever

During normal use of the hip joint (as in walking, running, and standing even momentarily on one leg), the force acting at the femoral head is the force of abductor muscle contraction (twice body weight) plus the force of total body weight, minus the weight of the supporting leg. This would suggest that a force of at least 2.5–3 × body weight passes through the femoral head when one stands on one leg. Since the act of standing on one leg is a brief but repetitive part of the gait cycle, it is easy to appreciate the considerable forces that pass through the hip joint and which increase in running and jumping activities.

Gait patterns

During the stance phase of the gait cycle (see Figure 16.5, p. 188), substantial forces are acting on the hip joint owing to the contraction of the abductor muscles. If these muscles are weak, or the fulcrum (the hip joint itself) is damaged, for whatever reason, the body

 Study tasks

• If necessary revise the principles of levers (Chapter 3), and highlight the details of the first-class lever that acts on the hip.
• Consider the biomechanical implications of the sex differences between male and female hips.

Clinical Note

The strength and repetitive nature of these forces combined perhaps with other factors that are detrimental to the physiology and biomechanics of the joint may predispose to degenerative changes, and in particular to osteoarthritis (osteoarthrosis) of the hip. Replacement surgery is now commonly undertaken in severe cases and has successfully improved the quality of life and life expectancy of sufferers in the majority of cases. For details of hip replacement techniques consult Williams (1995) and orthopaedic texts as appropriate.

Study tasks

- Consider what other physiological and biomechanical factors might contribute to degenerative changes in the hip joint.
- Consider the effects of using a walking stick on the affected side.

will not be supported during the stance phase. Either the body will sag on the unsupported side, producing what is called the 'Trendelenburg gait' (a waddling gait, if bilateral), or the body lurches towards the affected hip in an attempt to shorten the resistance arm of the lever, and uses gravity as a substitute for the hip flexors (known as an 'antalgic gait').

A walking stick on the unaffected side effectively reduces the resistance arm of the first-class lever, since it reduces the weight that the affected side has to support in the stance phase.

Chapter 15

The knee

Introduction

The knee is the largest synovial joint in the body. Like the shoulder, it should not be thought of as one single joint. It consists of the articulation between the femur and tibia (tibiofemoral joint), and the articulation between the patella and femur (patellofemoral joint), and from a functional point of view the superior tibiofibular joint should also be examined.

The knee is one of the most mechanically vulnerable areas, due to the forces and leverage acting on the joint both in terms of acute traumatic impacts and also in terms of chronic weightbearing problems. The knee will tend to be affected by dysfunction of the hip, ankle or foot, and it is important to consider the status of these joints in any analysis or diagnosis of knee problems.

The bones

In the knee, as in other joints, a preliminary examination of the bone features is crucial to an understanding of the mechanics of the joint. Particular attention should be paid to the articular surfaces of the femur and tibia in this respect.

The superior articular surface of the tibia addresses the femoral condyles as a horizontal surface would, and since the medial femoral condyle extends further distally than the lateral condyle, it results in the natural valgus of the knee (Figure 15.2)

Study tasks

- Highlight the features shown in Figure 15.1.
- Obtain matching bone specimens if possible, and identify the features shown.
- Place the femur on a flat surface with the shaft vertical and its anterior surface facing you. It is noticeable that the shaft deviates laterally at an angle of approximately 5–10° from the vertical. Also identify the two angles in the neck of the femur, referring if necessary to Chapter 14 (Figure 14.1(c)).

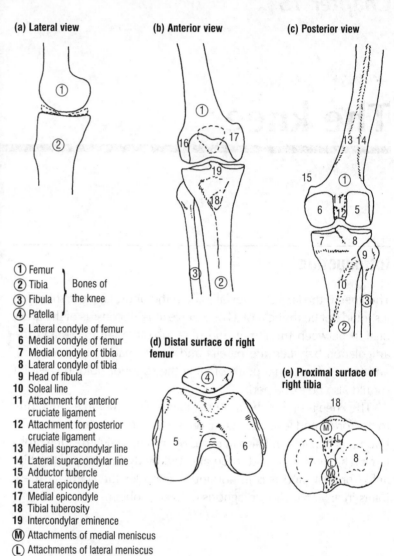

(a) Lateral view

(b) Anterior view

(c) Posterior view

(d) Distal surface of right femur

(e) Proximal surface of right tibia

① Femur ⎫
② Tibia ⎬ Bones of
③ Fibula ⎪ the knee
④ Patella ⎭
5 Lateral condyle of femur
6 Medial condyle of femur
7 Medial condyle of tibia
8 Lateral condyle of tibia
9 Head of fibula
10 Soleal line
11 Attachment for anterior cruciate ligament
12 Attachment for posterior cruciate ligament
13 Medial supracondylar line
14 Lateral supracondylar line
15 Adductor tubercle
16 Lateral epicondyle
17 Medial epicondyle
18 Tibial tuberosity
19 Intercondylar eminence
Ⓜ Attachments of medial meniscus
Ⓛ Attachments of lateral meniscus

Figure 15.1 The knee: bones and main bone features

The lateral condyle is more prominent on its anterior surface than the medial condyle. This helps to resist lateral displacement of the patella, which occurs due to the effect of the pull of the quadriceps muscle along the line of the femoral shaft, and the natural valgus of the knee. It also explains why the patella has asymmetrical medial and lateral facets. The oblique insertion into the patella of the vastus medialis part of the quadriceps muscle also opposes lateral displacement.

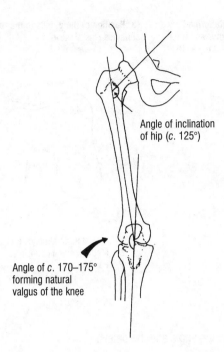

Angle of inclination
of hip (c. 125°)

Angle of c. 170–175°
forming natural
valgus of the knee

Figure 15.2 The natural valgus of the knee (anterior view)

(a) Genu valgum

(b) Genu varum

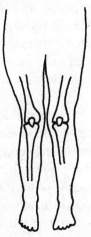
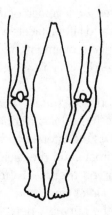

Figure 15.3 Deviations from the natural valgus of the knee

Clinical Note

An abnormal variation in this angle may result in 'genu valgum' or 'genu varum' (Figure 15.3). Compare this with cubitus valgus and varus in the elbow (p. 119)

(a) The 'Q' (quadriceps) angle formed between the shafts of the femur and tibia Normal values: Males, 13°; females, 18°

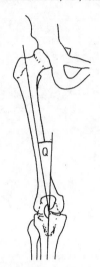

(b) Position of the patella in the femoral condyles (distal view)

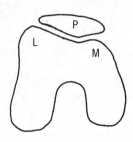

L   Lateral femoral condyle
M   Media femoral condyle
P   Patella

Figure 15.4 The quadriceps angle and the patella

The patella is the largest of the body's sesamoid bones, and lies embedded in the quadriceps tendon, keeping it 1–2 cm anterior to the tibiofemoral joint line. It rests in the grooved trochlear (pulley-shaped) surface of the anterior femoral condyles, and thereby increases the leverage of the quadriceps muscle. In addition to the medial and lateral facets which engage the femoral condyles during mid-range movements, there is a small so-called 'odd' facet on the medial side which makes contact only at approximately 135° of flexion.

In terms of surface contact with the tibia, the medial femoral condyle has a longer contact area in an anteroposterior direction than does the lateral condyle, and whereas the lateral condyle lies approximately in the sagittal plane, the medial condyle is noticeably curved (Figure 15.1(d)). The articular surface of the superior tibia is larger and more concave on the medial surface and this allows the curved wheel-like medial condyle to effect the various rotation movements that are integral to the function of the knee joint.

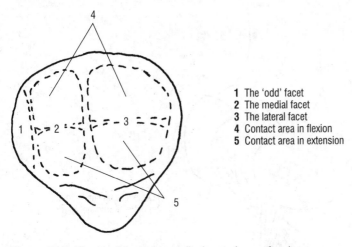

1 The 'odd' facet
2 The medial facet
3 The lateral facet
4 Contact area in flexion
5 Contact area in extension

Figure 15.5 Facets of the right patella (posterior surface)

Study tasks

- Try to obtain a bone specimen of a patella and identify the 'odd' facet.
- Consider why the lateral facet is more extensive than the medial facet.
- Consider why the anterior surface (not shown in Figure 15.5) is roughened.
- Shade the 'Q' angle in colour on Figure 15.4(a).

The ligaments and menisci

All joints are reinforced by the structural combination of ligaments, capsule and muscles. In the case of the bicondylar knee joint there is little inherent stability in the bone structure alone, and so the integrity of the joint is derived mainly from the soft tissues.

The main internal stability comes from the intracapsular cruciate ligaments and menisci. Lateral and medial stability is provided by the collateral ligaments, but, as in the shoulder, adequate support from the surrounding muscles is crucial for overall protection against injury.

The cruciate ligaments

The two cruciate ligaments (Latin *crux* = 'cross') lie inside the fibrous capsule of the knee (see below), but outside the synovial membrane (see also Figure 15.10(a)). Both ligaments tighten on full extension of the knee, and also act to restrain excessive rotation, particularly medial rotation of the tibia. The anterior cruciate prevents forward displacement of the tibia on the femur, and the posterior cruciate prevents backward displacement. The cruciates are also taut in full flexion; and at least one cruciate is taut in all positions, thereby providing vital tibiofemoral stability in spite of the limited congruity between the articular surfaces. Damage to the cruciate ligaments is particularly threatening to knee stability.

Study tasks

- Shade the anterior and posterior cruciate ligaments in separate colours.
- Locate their attachment points on bone specimens and on the diagrams of the superior surface of the tibia (Figure 15.1(e)).
- Simulate the cruciate ligaments of your right knee by pointing your right index finger towards your left foot (feet together). Now point your left index finger towards your right heel. Bring the two fingers together so that they form a cross. You now have the approximate arrangement of the cruciate ligaments of the right knee, with your right finger representing the anterior cruciate, and the left the posterior cruciate ligament.
- Palpate the role of the cruciate ligaments by placing a colleague supine with knees flexed – their feet should rest on the chosen surface. Place your hands around each leg in turn, close to the knee joint, and pull the tibia in an anterior direction, then push in a posterior direction. You should feel the resistance of the cruciate ligaments.
- Make sure that you understand which cruciate ligaments prevents further movement in each direction.

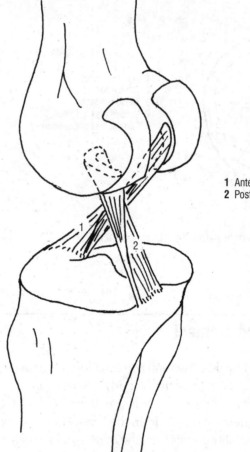

1 Anterior cruciate ligament
2 Posterior cruciate ligament

Figure 15.6 The cruciate ligaments of the right knee (posteromedial oblique view) – opened out

The collateral ligaments

The collateral ligaments provide stability to both the medial and lateral aspects of the joint and are palpable extracapsular structures. The medial (tibial) collateral ligament slants anteriorly from the femur to the tibia, and gains some protection at its tibial attachment from the overlying tendons of the sartorius, gracilis and semitendinosus muscles (Figure 15.9(a)). It tightens on full extension of the knee and opposes valgum forces (see Figure 15.3(a)) and lateral rotation of the tibia. Its deep fibres are connected to the edge of the medial meniscus.

The lateral (fibular) collateral ligament is shorter, more cord-like, and slants in a posterior direction from the lateral epicondyle of the femur towards the head of the fibula, where it is surrounded, and partly attached to, the tendon of the biceps femoris muscle. It is not attached to the lateral meniscus, and is separated from it by the tendon of the popliteus muscle.

The lateral collateral ligament tightens on full extension, opposes varum forces (see Figure 15.3(b)) and, like the medial collateral ligament, resists lateral rotation of the tibia.

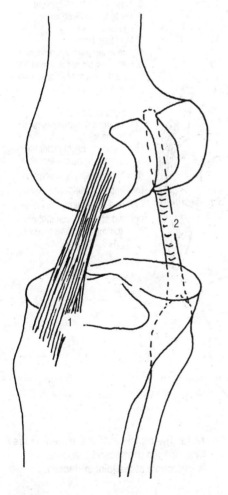

1 Medial (tibial) collateral ligament
2 Lateral (fibular) collateral ligament

Figure 15.7 The collateral ligaments of the right knee (posteromedial oblique view) – opened out

Study tasks

- Highlight the details shown, including the functions of the menisci listed in Figure 15.8.

The menisci

The menisci or semi-lunar cartilages are crescent-shaped wedges of fibrocartilage which are attached by their tips (horns) to the inter-condylar area of the tibia, and lie inside the synovial membrane (Figure 15.10(a)). Anteriorly, they are connected to each other by a transverse ligament. The medial meniscus has attachments to the

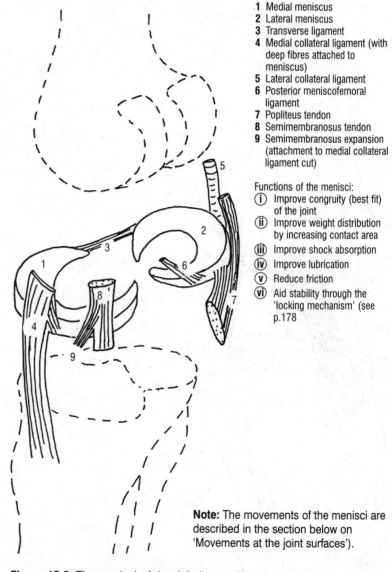

1 Medial meniscus
2 Lateral meniscus
3 Transverse ligament
4 Medial collateral ligament (with deep fibres attached to meniscus)
5 Lateral collateral ligament
6 Posterior meniscofemoral ligament
7 Popliteus tendon
8 Semimembranosus tendon
9 Semimembranosus expansion (attachment to medial collateral ligament cut)

Functions of the menisci:
(i) Improve congruity (best fit) of the joint
(ii) Improve weight distribution by increasing contact area
(iii) Improve shock absorption
(iv) Improve lubrication
(v) Reduce friction
(vi) Aid stability through the 'locking mechanism' (see p.178)

Note: The movements of the menisci are described in the section below on 'Movements at the joint surfaces').

Figure 15.8 The menisci of the right knee with selected attachments (posteromedial oblique view) – opened out

fibrous capsule and the deep fibres of the medial collateral ligament, which are in turn connected to an expansion from the semimembranosus muscle. Inferiorly, the menisci are attached to the tibia by capsular fibres which are often referred to as the 'coronary ligaments'. The lateral meniscus is more mobile, partly due to its more circular shape, but also due to its attachment to the tendon of the popliteus muscle, which assists posterior movement during knee flexion. The posterior meniscofemoral ligament links the posterior part of the meniscus to the medial femoral condyle.

In spite of their peripheral attachments, the menisci retain sufficient mobility to allow them to adapt to knee movements. They are thicker peripherally, and vascularised by the capsule and synovial membrane, but are significantly avascular on their thinner inner margins – which has clinical implications for the healing of damaged or torn menisci.

The capsule and related structures

A fibrous capsule surrounds the tibiofemoral joint, but the presence of the patella and the internal structures of the knee make it a complex structure. It is attached to the margins of the hyaline cartilage that covers the articular surfaces, and it surrounds the sides and posterior aspect of the joint, but is deficient anteriorly because of the presence of the patella and to allow access for the important suprapatellar bursa (Figure 15.10(b)). The medial and lateral sides of the patella, in effect, make good this deficiency by receiving fibres (the lateral and medial retinacula) from, respectively, the vastus lateralis and medialis portions of the quadriceps muscle. These fibres blend with the capsule and extend as far as the collateral ligaments on each side. They also help to stabilise the patella when the knee is flexed and loaded. The quadriceps muscle above and its tendinous attachment point, the ligamentum patellae, below also make good the anterior deficiency in the capsule. Posteriorly, the tendon of the popliteus emerges from the capsule, but is reinforced by the arcuate and oblique popliteal ligaments, the latter of which is an expansion of the semimembranosus tendon.

(a) Anterior view

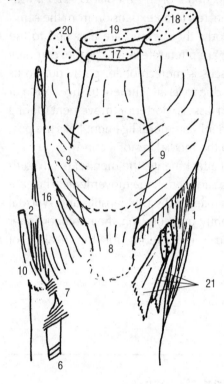

(b) Posterior view

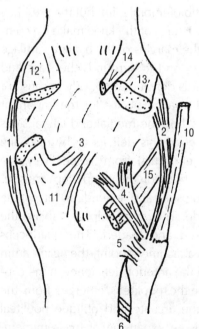

1 Medial collateral ligament
2 Lateral collateral ligament
3 Oblique popliteal ligament
4 Arcuate ligament
5 Posterior ligament of superior
   tibiofibular joint
6 Interosseous membrane (part)
7 Anterior ligament of superior
   tibiofibular joint
8 Ligamentum patellae
9 Patellar retinacula
10 Biceps femoris tendon
11 Semimembranosus tendon
12 Medial head of gastrocnemius
13 Lateral head of gastrocnemius
14 Plantaris tendon
15 Popliteus tendon
16 Iliotibial tract
17 Rectus femoris
18 Vastus medialis
19 Vastus intermedius
20 Vastus lateralis
21 Tendons of sartorius, gracilis
   and semitendinosus
   (superficial to deep)

Figure 15.9 Various supporting structures of the right knee

The synovial membrane, bursae and fat pads

(a) Superior articular surface of right tibia (oblique view)

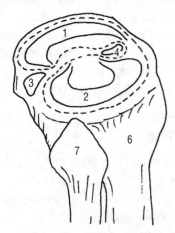

1 Medial meniscus
2 Lateral meniscus
3 Attachment for posterior cruciate ligament
4 Attachment for anterior cruciate ligament
5 Femur
6 Tibia
7 Fibula
8 Patella
9 Quadriceps tendon
10 Prepatellar bursa
11 Ligamentum patellae
12 Deep infrapatellar bursa
13 Superficial infrapatellar bursa
14 Articularis genus muscle*
15 Infrapatellar fat pad
16 Suprapatellar bursa

--·~·-·-- Synovial membrane

* A small muscle that suspends the suprapatellar bursa and attaches to the anterior shaft of the femur

(b) Knee (sagittal view)

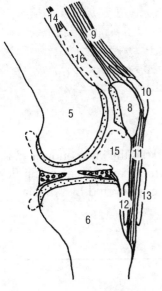

Figure 15.10 The synovial membrane and selected structures of the right knee

The synovial membrane lines the capsule, and emerges from the deficient anterior part of the capsule to form the suprapatellar bursa. The synovium follows a convoluted path inside the knee, particularly anteriorly, where it encloses the menisci but does not include the cruciate ligaments, which are intracapsular but extrasynovial (Figure 15.10(a)). Posteriorly, it is moulded by tendons into recesses that form the semimembranosus and gastrocnemius bursae, and the subpopliteal recess; all of which communicate with the main synovial cavity. There are also a number of non-communicating bursae that are clinically important, such as the 'pes anserinus' bursa (literally 'goose foot shaped') which lies between the tendons of the sartorius, gracilis and semitendinosus and the medial collateral ligament. Others are shown in Figure 15.10(b), but for a more complete list the reader should consult Williams (1995).

Clinical Note

Synovitis of the knee joint, for whatever reason, will lead to effusion (excessive synovial fluid) within the capsule and suprapatellar bursa. This gives rise to the condition popularly known as 'water on the knee'. Inflammation of the individual bursae causes localised swelling and bursitis, with occasionally colourful descriptions. For example, prepatellar bursitis is termed 'housemaid's knee' (caused by leaning forwards on the knees), and infrapatellar bursitis is termed 'parson's knee' (perhaps caused by rocking back into a position of prayer). If 'cleanliness is next to godliness' we should be careful when we scrub and pray.

A substantial infrapatellar fat pad fills the intercondylar space of the femur and moves across the trochlear surface during flexion and extension. The articular surface is covered by synovial membrane which aids lubrication.

Movements and muscles

The active movements of the knee are flexion and extension, and rotation if the knee is flexed or partially flexed. The tibiofemoral joint is the bicondylar mechanism that allows the length of the lower limb to be varied either slowly or rapidly, like a spring. This permits shock absorption and leverage, with an element of rotation which facilitates changes in direction of gait (for example, during sport or dancing).

The patellofemoral joint is especially important in flexion/extension, increasing the mechanical advantage of the quadriceps tendon.

The role of the superior tibiofibular joint is passive: it is not actively involved, but moves permissively with the main knee and ankle movements.

(a) Flexion of the knee

(0–140°)

(0–120°)

Movement produced by:
• biceps femoris
• semitendinosus
• semimembranosus

assisted by:
• gracilis
• sartorius
• popliteus
• gastrocnemius ⎫
• plantaris ⎭ when the foot is 'grounded'

(b) Extension of the knee

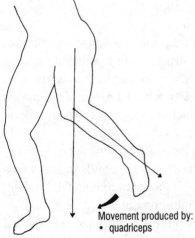

Movement produced by:
• quadriceps

assisted by:
• tensor fasciae latae

c. 5° hyperextension available on passive stretch

(c) Medial rotation of the knee (0–30°)

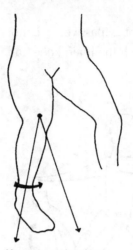

(d) Lateral rotation of the knee (0–40°)

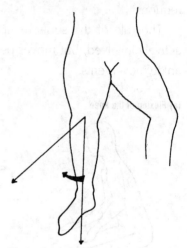

Movement produced by:
- popliteus
- semimembranosus
- semitendinosus

assisted by:
- sartorius
- gracilis

Movement produced by:
- biceps femoris

Figure 15.11 Movements of the knee

Movements at the joint surfaces

In the case of a complex joint such as the knee, it is important to understand the movements that are taking place at the contact points of the joint surfaces, as well as to appreciate the gross movements of flexion, extension and rotation.

During the movements of flexion, extension and rotation, the femoral condyles roll and slide on the superior surface of the tibia in a manner that reflects the bone structure of the articular surfaces. (See Chapter 3, pp. 16–17 for explanations of the terms 'roll' and 'slide'). 'Roll' may be compared with a wheel rolling along a road and is more stable than 'slide', which is similar to a skid and therefore less stable.

From a position of extension (the anatomical position) to about 20° flexion (a range of movement that corresponds to the stance phase in the gait cycle (p. 188), roll is more predominant than slide. At this stage, stability is the main requirement.

After approximately 20° flexion the radius of the femoral condyles reduces posteriorly, the ligaments between the tibia and femur are more relaxed, and the tibiofemoral joint is relatively looser. Slide is more important than roll after 20° flexion, and overall stability is greatly influenced by the strength of the quadriceps muscle and the integrity of the patella, which is forced against the femoral condyles.

The menisci also move, and in effect distort, on the superior surface of the tibia to which they are attached. Students are often confused by the movements of the menisci, but may be helped if they remember the 'golden rule' that the menisci always follow the movements of the femoral condyles. In flexion the condyles move posteriorly, and cause the menisci to distort in this direction; whereas in extension the movements are anterior. The lateral meniscus is more mobile, and undergoes greater distortion due to its attachment points (horns) being closer together. In flexion it is also pulled posteriorly by its attachment to the popliteus tendon, and in extension it is drawn forwards (as is the medial meniscus) by tension in the transverse ligament, which is in turn pulled by fibres attached to the patellar ligament (meniscopatellar fibres). The posterior horn of the lateral meniscus is also drawn forwards in extension by the meniscofemoral ligament as the posterior cruciate ligament tightens.

By contrast, the medial meniscus is less mobile. Its periphery is attached to the joint capsule and medial collateral ligament. It may also be pulled posteriorly in flexion by the attachment of an expansion of the semimembranosus muscle to the medial collateral ligament.

(a) Flexion

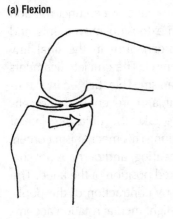

(Sagittal view)

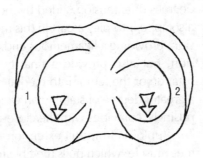

(Superior view of right tibia)

Study tasks

- Highlight the details shown in Figure 15.12, and colour the arrows showing the difference between flexion and extension movements.

Clinical Note

The greater mobility of the lateral meniscus means that it is less vulnerable to injury than the medial meniscus, which is frequently involved in injuries to the medial collateral, and the anterior cruciate ligament – a syndrome that is sometimes referred to as the 'unhappy triad of O'Donoghue' (O'Donoghue, 1950).

(b) Extension

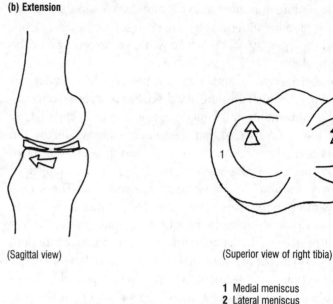

(Sagittal view)

(Superior view of right tibia)

1 Medial meniscus
2 Lateral meniscus

Note: Arrows show direction of meniscal movements only.

Figure 15.12 Movements of the menisci in flexion and extension of the knee

Further important structural features of the femoral condyles are the curved shape and oblique orientation of the medial femoral condyle, and the more sagittally directed lateral condyle. This ensures that the medial surface of the tibia rotates during flexion/extension. This is a passive, automatic or conjunct rotation (spin), which accompanies all flexion/extension movements and acts as a 'screw home' or locking mechanism in the final few degrees of extension, aided by the menisci. The cruciate ligaments also appear to play a part in this mechanism. They tighten in extension, providing a ligamentous end point, and prevent further extension, but also provide a final twist to take the tibia into lateral rotation or the femur into medial rotation. This mechanism confers a greater degree of stability on weightbearing, and perhaps not surprisingly, coincides with the close-packed position of the knee. The knee unlocks from full extension (0°) by contraction of the popliteus muscle, which pulls the tibia into slight medial rotation accompanied by lateral rotation of the femur.

(a) Anterior view of flexed right knee (90°)

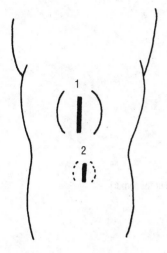

(b) Anterior view of fully extended right knee

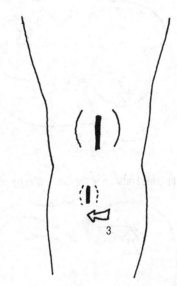

1 Vertical line marked through midpoint of patella
2 Vertical line marked through midpoint of tibial tuberosity
3 Tibial tuberosity appears to move laterally

Figure 15.13 Helfet's test

 Study task

• Observe conjunct rotation in extension of the knee by performing 'Helfet's test' on a colleague. To do this, seat them with knees and hips flexed to 90° and knees dangling free. Mark the skin over the medial and lateral borders of the patellae. Then mark two vertical lines: one down the midline of the patellae, and one rising vertically from the tibial tuberosities (Figure 15.13(a)). The tibial tuberosities should be in line with the medial half of the patellae in flexion, but if you ask your colleague to extend their knees fully, you should observe lateral movement of the lines over the tibial tuberosities in normal knees (Figure 15.13(b)).

This 'conjunct' rotation which accompanies flexion/extension should not be confused with the active 'adjunct' rotatory movements, medial and lateral, which are an important part of normal knee function. The curved structure of the medial femoral condyle facilitates rotation; and once again the menisci follow the movements of both femoral condyles. In this context, it is important to bear in mind that when the femur rotates laterally the tibia undergoes relative medial rotation and vice versa. Thus, during medial rotation of the femur (lateral rotation of the knee or tibia) the medial meniscus is drawn posteriorly and the lateral meniscus moves anteriorly. In contrast, during medial rotation of the knee (tibia), the medial meniscus is drawn anteriorly and the lateral meniscus posteriorly, by the lateral rotation (spin) of the femur.

Study tasks

- Shade the arrows in colour, showing the movements of the menisci in medial and lateral rotation of the knee.
- Ensure that you understand the relative movements of the tibia, femur and menisci in rotation.

Study tasks

- Revise the knee joint by carrying out a passive supine clinical examination of a colleague's knees, using a treatment table (plinth), noting any differences between the two knees. Adopt the following protocol:
  - *Observation*: Review all surface anatomy features using anatomy resources as necessary, ensuring that you can identify all visible features. Try to identify any discrepancies that you notice (scars, swelling, etc.). Check whether the knee remains straight (0° extension) or remains in a degree of flexion.
  - *Palpation*: Note any swelling or temperature differences that may indicate inflammation. If you suspect that your colleague does not have a healthy knee, seek qualified guidance before proceeding. Identify, where possible all soft tissue and bone structures that have been studied in this chapter.
  - *Active movements*: Ask your colleague to flex, extend and, in flexion, medially and laterally rotate each knee in turn. Note any differences or discomfort. This may be done on or off weightbearing, but the latter puts
    (cont. opposite)

(a) Lateral rotation of the knee (tibia) (medial rotation of the femur)

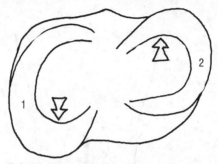

(Superior view of right tibia)

(b) Medial rotation of the knee (tibia) (lateral rotation of the femur)

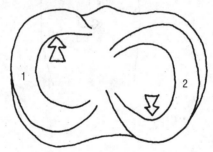

1 Medial meniscus
2 Lateral meniscus

Note: Arrows show direction of meniscal movements only.

(Superior view of right tibia)

Figure 15.14 Movements of the menisci in rotation movements of the knee

Note: It is conventional to refer to the direction of knee rotation as being the direction in which the tibia rotates. Thus, medial rotation of the knee is the equivalent of medial rotation of the tibia. Rotation of the femur will be lateral in this case.

Clinical Note

If the menisci fail to move with the femoral condyles they may be torn, split or become detached. This may happen during a violent extension movement (for example, a misplaced kick) or during a sudden rotatory movement during weightbearing and flexion. The damaged meniscus may fail to follow normal movements, resulting in locking of the knee in flexion. Damage to the synovial membrane will tend to cause massive synovial effusion; and because of the avascular nature of the inner menisci, healing can be expected to take place only in the peripheral parts.

Instant centre analysis

The role of instant centre analysis was discussed in Chapter 3, and can provide a valuable means of assessing whether or not a tibiofemoral joint is functioning properly.

In order to produce an instant centre pathway of the tibiofemoral joint, a number of lateral radiographs (X-rays/roentgenograms) are taken from a position of extension into flexion at 10° intervals. Two points on the tibia that can be easily identified on all radiographs are chosen, and then lines are drawn to connect the paired points. A perpendicular may then be drawn from the line joining each pair, and the point of intersection gives the instant centre. If this process is repeated a number of times, the instant centres may be seen to form a semicircular pathway in a healthy knee that rolls and slides in a normal manner (Figure 15.15(b)). Any derangement of normal knee mechanics (for example, in a meniscal tear) will tend to produce an erratic instant centre pathway (Figure 15.15(a)).

(a) Erratic pathway of instant centres in a dysfunctional knee

(b) Smooth pathway of instant centres in a normal knee

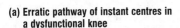

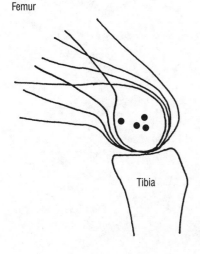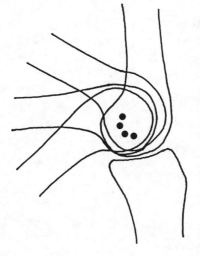

Figure 15.15 Hypothetical application of instant centre analysis

Study tasks (cont.)

considerably more force through the joint and may cause pain or discomfort if the joint is dysfunctional. For this reason, the movement should be performed carefully.

- *Passive movements*: Commence by inspecting the position of the patellae. Palpate the tibiofemoral joint line. (The inferior edge (pole) of the patella should be in line with this.)
- Lift each extended leg in turn. The knee should reveal a few degrees of additional (passive) extension.
- Gently flex each knee in turn by supporting the heel with one hand and palpating the tibiofemoral joint line with the other. Take the knee into full flexion, noting, and stopping at, any point of discomfort.
- Return the knee to 90° flexion and, with your supporting hand under the heel, guide your colleague's knee into medial and lateral rotation. Note any differences in the range of movement. Support and lift each knee with both hands in turn; the lower leg should be comfortably tucked under your arm (at the axilla).
- From this position you can check lateral shift (gapping), which is an accessory knee movement and not under active control. If the knee is semi-flexed, slight but not excessive gapping should be palpable (lateral and medial shift, or translation), which tests the integrity of the collateral ligaments.

(cont. p. 182)

Study tasks (cont.)

- Check the integrity of the cruciate ligaments by placing your colleague's flexed knee at 90° with their foot resting on the plinth. Place both your hands around the upper tibia below the tibiofemoral joint with your thumbs resting on the tibial tuberosity. Gently pull the leg towards you, and you should detect a slight but not excessive shift. This is known as the 'anterior drawer test' and examines the integrity of the anterior cruciate ligament. Follow this by pushing the tibia posteriorly to test the posterior cruciate ligament (the 'posterior drawer test').
- Finally, palpate the movement that can be felt at the superior tibiofibular joint (Figure 15.9). This is a synovial plane joint that allows slight lateral rotation of the fibula during dorsiflexion of the ankle joint (Williams, 1995). The head of the fibula may be gently grasped and moved passively, very slightly, in an anterior and posterior direction.
- Consider also which tissues limit each passive movement that you performed earlier.

Clinical Note

The knee, like the hip, is another weightbearing area of the body which commonly suffers from degenerative changes. Any alteration in the normal biomechanics of the joint surfaces (including the patellar surfaces) may initiate degenerative changes which may develop into osteoarthritis. Advanced cases of osteoarthritis are now often successfully treated by knee replacement surgery. For further details consult orthopaedic texts as appropriate, but an introduction to the subject may be found in Palastanga et al. (1994) and Williams (1995).

Study tasks

- Consider the factors that may commonly alter the biomechanics of the knee joints.
- Consider any other factors that might predispose to degenerative changes in the knee joints.

Chapter 16

The ankle

Introduction

The function of the ankle is both load bearing and locomotive/propulsive. It is somewhat like a mortise and tenon joint in carpentry, with the talus bone as the tenon, and the surrounding mortise formed by the distal parts of the tibia and fibula and their connecting ligaments. It is therefore valid to refer to the ankle joint as the 'talocrural' joint, which means 'of the ankle and lower leg bones'. The complex movements of the foot which allow adaptation to changes in ground surface occur below this at the subtalar joints and the joints of the forefoot (see Chapter 17).

The talocrural joint itself is basically a uni-axial hinge joint. The articular surfaces are to some extent saddle shaped; but the essential movements are those of plantar flexion and dorsiflexion (see Figure 16.3), which are so important in the propulsive actions of walking and running (the gait cycle).

The bones

(a) Bones and bone features of the right ankle (posterior view)

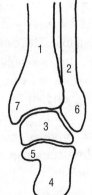

1 Tibia (distal)
2 Fibula (distal)
3 Talus
4 Calcaneus
5 Sustentaculum tali ('ledge for the talus')
6 Lateral malleolus
7 Medial malleolus

Study tasks

- Try to obtain matching specimens of a tibia, fibula, talus and, if possible, calcaneus. If this is difficult refer to detailed illustrations of these bones; see, for example, Williams (1995).
- Observe the greater anterior width of the superior surface of the talus. Notice also that the lateral articular surface of the talus lies at approximately 90° to the essentially transverse axis of movement of the talocrural joint (on the bone specimens) whereas the medial surface does not.
- Test the axis of movement on your own ankle by placing your fingertips on your medial and lateral malleoli (ankles). Move your ankle from plantar flexion to dorsiflexion, and notice that the axis lies posterior to the frontal plane on the lateral side. The axis also appears to change slightly with ankle motion.
- If a specimen is available, move the bones into positions of plantar flexion and dorsiflexion. From these movements alone, try to decide which is the most stable and which is the least stable position for the ankle joint.

(b) Superior view of the right talus

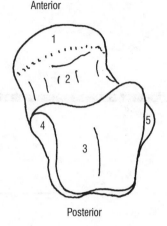

1 Head
2 Neck
3 Trochlear surface (for articulation with the tibia and fibula)
4 Facet for medial malleolus (tibia)
5 Facet for lateral malleolus (fibula)

Figure 16.1 The ankle: bones and bone features

The joint and ligaments

The inferior tibiofibular joint which connects the distal ends of the tibia and fibula is a fibrous joint, or syndesmosis. It is bound strongly by an interosseous ligament, and is continuous with the interosseous membrane above which connects the shafts of the tibia and fibula. The surfaces of the distal bones are slightly separated by an intrusion of synovial membrane which lines the capsule of the talocrural joint. Anteriorly the inferior tibiofibular joint is supported by the anterior tibiofibular ligament (Figure 16.2 (a)), and posteriorly by the posterior tibiofibular ligament (Figure 16.2 (b)). A deep portion of the latter forms the inferior transverse ligament (not shown), which is partly covered by hyaline cartilage, and forms part of the articular surface of the talocrural joint.

The talocrural joint is basically uni-axial, but the axis of plantar and dorsiflexion is dynamic. The capsule that surrounds the articular surfaces is relatively thin (for mobility), and the joint line is easily palpable as a slight hollow, anteriorly, with the ankle relaxed. The ankle joint is highly dependent on the strong collateral ligaments (Figure 16.2 (c–d)), as well as the muscles of dorsiflexion, plantar flexion and those that produce inversion/eversion movements in the subtalar joints below (see Chapter 17).

The weakness of the talocrural joint lies in the almost dome-like pivotal role of the talus, which has no muscle attachments and is therefore highly dependent on ligamentous support to prevent anterior or posterior displacement of the leg on the talus, excessive internal/external rotation or inversion/eversion strains. The medial collateral (deltoid) ligament is particularly strong, and resists medial collapse of the ankle into eversion, as well as internal/external rotation of the tibia. Rather more vulnerable are the individual parts of the lateral collateral ligament, referred to separately as the 'anterior talofibular', 'posterior talofibular' and 'calcaneofibular' ligaments (Figure 16.2 (d)).

Of these, the anterior talofibular ligament becomes vertically oriented and taut in plantar flexion: it then acts to restrict inversion in plantar flexion, but is vulnerable in this function.

The calcaneofibular ligament tightens in dorsiflexion, and also acts to restrict inversion strain; the posterior talofibular ligament resists posterior displacement and internal rotation.

(a) Anterior view

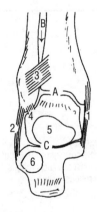

1 Deltoid (medial) ligament
2 Calcaneofibular ligament (part of lateral ligament)
3 Anterior tibiofibular ligament
4 Anterior talofibular ligament
5 Head of talus
6 Facet for cuboid on calcaneus

A Tibiotalar joint (hinge)
B Inferior tibiofibular joint (syndesmosis)
C Subtalar joint* (anterior, or talocalcaneonavicular part; multi-axial)

* This joint permits eversion/inversion of the foot (see p. 194)

(b) Posterior view

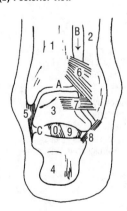

1 Tibia
2 Fibula
3 Talus
4 Calcaneus
5 Deltoid (medial) ligament
6 Posterior tibiofibular ligament
7 Posterior talofibular ligament
8 Calcaneofibular ligament (part of lateral ligament)
9 Cervical ligament (part)
10 Interosseous talocalcanean ligament

A Tibiotalar joint (hinge)
B Inferior tibiofibular joint (syndesmosis)
C Subtalar joint* (posterior, or talocalcanean part) (modified multi-axial)
* This joint permits eversion/inversion of the foot (see p. 194)

📖 ***Study tasks***

• Shade the ligaments in colour and highlight the details given.
• Consider why it is important that the inferior tibiofibular joint is a syndesmosis, but retains some movement.
• Palpate the anterior joint line of the ankle, noticing that it forms a palpable hollow in the early stages of plantar flexion, a hollow that is obliterated in dorsiflexion. Consider why this should be so.

(c) Medial view

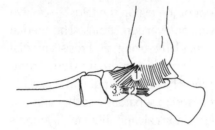

1 Medial collateral (deltoid) ligament
2 Plantar calcaneonavicular (spring) ligament
3 Tuberosity of navicular

(d) Lateral view

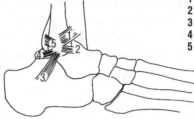

1 Anterior tibiofibular ligament   ⎫ The lateral
2 Anterior talofibular ligament   ⎬ collateral
3 Calcaneofibular ligament   ⎭ ligament
4 Posterior talofibular ligament
5 Posterior tibiofibular ligament

Figure 16.2 Ligaments and joints surrounding the ankle

Movements and muscles

(a) Dorsiflexion of the ankle (0–15°)

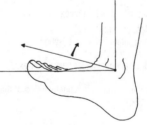

Movement produced by:
• tibialis anterior

assisted by:
• extensor digitorum longus
• extensor hallucis longus
• peroneus tertius

(b) Plantar flexion of the ankle (0–55°)

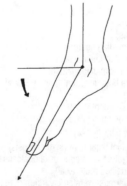

Movement produced by:
• gastrocnemius
• soleus

assisted by:
• plantaris
• tibialis posterior
• flexor hallucis longus
• flexor digitorum longus

Figure 16.3 Movements fo the ankle joint

Biomechanics of the ankle

As the ankle moves from plantar flexion to dorsiflexion the talus slides back in the mortise of the talocrural joint. Examination of the superior articular surface of the talus reveals that the lateral surface is longer than the medial surface in an antero-posterior direction. This results in the lateral malleolus travelling a greater distance over the talus than the medial malleolus, and the result is what may be described as an 'impure' swing involving a degree of spin or rotation. The trochlear (superior) surface of the talus is sometimes described as a cone with an apex located medially which rotates about its long axis within the mortise of the joint (Kessler and Hertling, 1990).

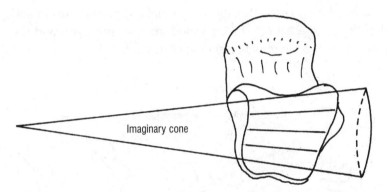

Imaginary cone

Figure 16.4 The conical disposition of the trochlear surface of the talus (right/superior view)

This configuration explains why the malleoli do not separate significantly in dorsiflexion in spite of the difference in anterior and posterior widths of the trochlea of the talus. It also explains why the axis of plantar/dorsiflexion changes slightly during the movements – a feature that may be observed by palpating the malleoli (see study task p. 184). As in the case of the tibiofemoral joint of the knee, instant centre analysis has provided valuable insight into the differences between normal and abnormal movements in the ankle joint (Sammarco *et al.*, 1973). If movements are measured from plantar flexion to dorsiflexion, instant centre analysis reveals that in a normal ankle the joint surfaces are relatively distracted in plantar flexion. The surfaces then show a tendency to slide before ultimately showing compression in dorsiflexion at the end of range. This contrast between distraction and compression may enhance synovial lubrication of the joint surfaces.

Clinical Note

The position of plantar flexion is less stable than that of dorsiflexion owing to the narrower width of the posterior surface of the superior talus. Stability is provided by the vertical alignment of the anterior parts of the collateral ligaments; but the ankle is vulnerable under weightbearing conditions, and if the foot collapses into inversion with plantar flexion the anterior talofibular ligament is particularly at risk. Inversion strains to the lateral part of the ankle are common.

In contrast, the same study suggested that abnormal ankle joints behave in an erratic manner, alternating somewhat unpredictably between distraction and compression.

Another feature of the normal talocrural joint is that load bearing is relatively well spread; the normal ankle seeming to be spared the degenerative changes that might be expected in a joint that seems to act as a dome or focal point in weightbearing. Ramsey and Hamilton (1976) have shown that a slight lateral shift of the talus (often a sequel to fractures and ligamentous injury) can concentrate weightbearing on the medial trochlea, which might precipitate degenerative changes due to altered mechanics.

The role of the ankle in posture and gait

It should be apparent that the movements of plantar flexion and dorsiflexion are actually incorporated into the gait cycle and the maintenance of an upright standing posture.

Study task

- Consider at which stages of the gait cycle the muscles of plantar flexion and dorsiflexion are most active.

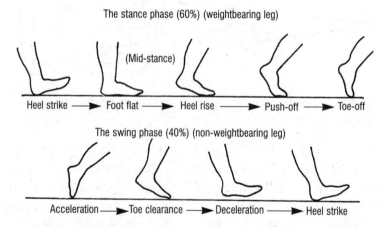

Figure 16.5 The gait cycle

In a normal standing position, approximately 50% body weight is distributed through each foot, with a centre-of-gravity line (Figure 6.2(c)) passing slightly anterior to the transverse axis of the ankle joint. This means that the body has a slight tendency to topple forwards, which is countered by downward pressure of the feet (plantar flexion at the ankle joint) caused by contraction of the plantar flexor muscles (particularly the soleus).

Clinical Note

Venous return in the lower limb is greatly aided by the alternate contraction and relaxation of the calf muscles (the so-called 'calf pump' mechanism). Prolonged standing will inhibit this mechanism and may predispose to syncope (fainting). Individuals with a tendency to anterior weightbearing will tend to rely on the continual recruitment of the plantar flexor muscles to prevent themselves falling forwards. The resultant hypertonia of these muscles will also inhibit the calf pump mechanism, and together with other factors may in the long term become a predisposition to varicose veins.

If the body is raised on tiptoe, the force that passes through the tendo-calcaneus (Achilles tendon) is approximately 1.2 times body weight, with the total joint reaction force passing through the ankle rising to 2.1 times body weight (Frankel and Nordin, 1989).

Lambert (1971) has also demonstrated that the fibula carries up to one-sixth of this load, contrary to the widely held belief that the bone is non-weightbearing.

During a normal gait cycle (Figure 16.5), the lower limb moves through space in a sagittal plane. The axis of motion in the ankle joint is not perpendicular to the sagittal plane and deviates laterally by about 25°, though this is compensated by the subtalar joints below (see Chapter 17). At heel strike the ankle should begin to plantar flex. This increases to the point of 'foot flat', but changes to dorsiflexion at mid-stance. At the end of the stance phase, the ankle returns to plantar flexion to allow 'push-off/toe-off'.

In the middle of the 'swing phase' (non-weightbearing leg), the motion changes to dorsiflexion, reverting to plantar flexion once again at heel strike. The height of shoe heel greatly influences the range of movement permitted in each foot.

Note: For a more detailed analysis of the functional anatomy of the gait cycle, including reference to other joints of the lower limb, see for, example, Kessler and Hertling (1990).

Study tasks

- Consider the implications of wearing high-heeled shoes on the calf pump mechanism and the muscles of plantar flexion

Clinical Note

The use of a heel lift in a shoe is often considered appropriate in the treatment of repetitive strain injuries to the tendo-calcaneus, or of calf muscle tears.

Study tasks

- Consider the significance of the following factors to a runner who wishes to prevent calf injuries: (i) choice of running shoe; (ii) warming-up and stretching exercises; (iii) change in running speed; (iv) change in slope of running surface; (v) change in surface from hard to soft ground; (vi) running style. Note here that sprinters favour landing on the ball of the foot in order to accelerate push-off/toe off. Consider why this particular style may be inadvisable when running longer distances.
- Consider reasons why the occurrence of osteoarthritis of the ankle joint is less prevalent than in the hip and knee joints.

Chapter 17

The foot

Introduction

The foot is the body's primary contact point with the ground; and since the ground is often uneven, the foot must be adaptable. It supports body weight and provides a strong platform and leverage for body movement. It has an inherently marvellous structure which contains 26 bones (plus two sesamoids), and 57 joints that possess flexibility; yet it also converts into a relatively rigid structure which allows a ballet dancer to stand on points.

It is important to realise that the entire body is affected by the mechanics of the foot, through its influence on the ankle, hip, knee, pelvis and vertebral column.

In fact it is reasonable to regard the foot as the foundation upon which everything else stands, and that problems that alter the mechanics of the foot will influence body posture and vertebral structures. The correction of many neuromusculoskeletal problems may usefully and properly begin with analysis of the foot.

The bones

(a) Dorsal view

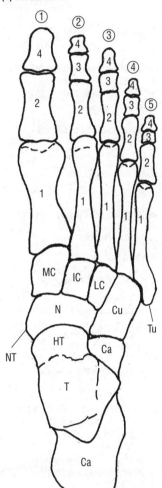

(b) Plantar view

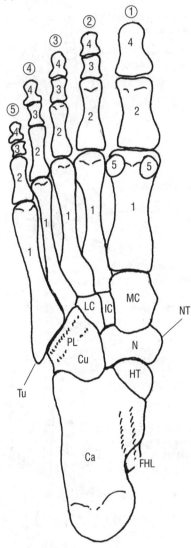

> **Study tasks**
>
> - Highlight the names of the bones and bone features shown.
> - Obtain a bone specimen of the entire foot and examine the features shown on the diagrams.
> - Palpate the navicular tuberosity, the tuberosity on the base of the fifth metacarpal, and the head of the talus.

①–⑤ Anatomical designation of digits
1 Metatarsals
2 Proximal phalanges (sing. phalanx)
3 Middle phalanges
4 Distal phalanges
5 Sesamoid

T	Talus
Ca	Calcaneus
N	Navicular
Cu	Cuboid
MC	Medial cuneiform
IC	Intermediate cuneiform
LC	Lateral cuneiform
NT	Navicular tuberosity
HT	Head of the talus
Tu	Tuberosity of fifth metatarsal
PL	Groove for peroneus longus tendon
FHL	Groove for flexor hallucis longus tendon

Figure 17.1 The foot: bones and selected bone features

The joints and ligaments: introduction

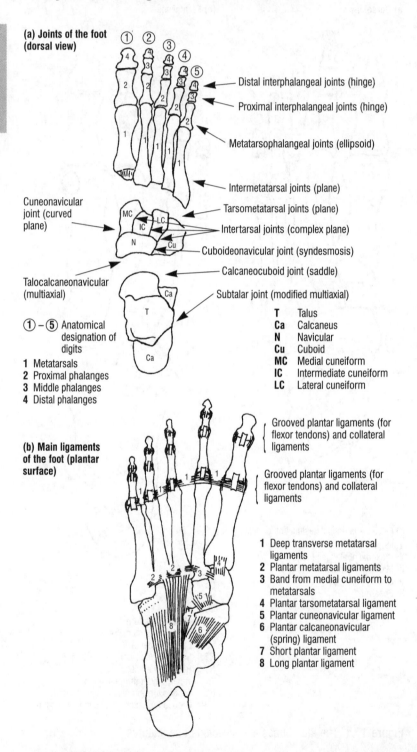

(a) Joints of the foot (dorsal view)

Distal interphalangeal joints (hinge)

Proximal interphalangeal joints (hinge)

Metatarsophalangeal joints (ellipsoid)

Intermetatarsal joints (plane)

Cuneonavicular joint (curved plane)

Tarsometatarsal joints (plane)

Intertarsal joints (complex plane)

Cuboideonavicular joint (syndesmosis)

Calcaneocuboid joint (saddle)

Talocalcaneonavicular (multiaxial)

Subtalar joint (modified multiaxial)

① – ⑤ Anatomical designation of digits
1 Metatarsals
2 Proximal phalanges
3 Middle phalanges
4 Distal phalanges

T Talus
Ca Calcaneus
N Navicular
Cu Cuboid
MC Medial cuneiform
IC Intermediate cuneiform
LC Lateral cuneiform

(b) Main ligaments of the foot (plantar surface)

Grooved plantar ligaments (for flexor tendons) and collateral ligaments

Grooved plantar ligaments (for flexor tendons) and collateral ligaments

1 Deep transverse metatarsal ligaments
2 Plantar metatarsal ligaments
3 Band from medial cuneiform to metatarsals
4 Plantar tarsometatarsal ligament
5 Plantar cuneonavicular ligament
6 Plantar calcaneonavicular (spring) ligament
7 Short plantar ligament
8 Long plantar ligament

(c) Retinacula (medial view)

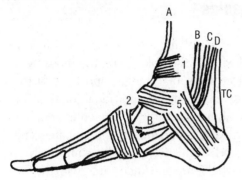

1 Superior extensor retinaculum
2 Inferior extensor retinaculum
3 Superior peroneal retinaculum
4 Inferior peroneal retinaculum
5 Flexor retinaculum

Retinacula (lateral view)

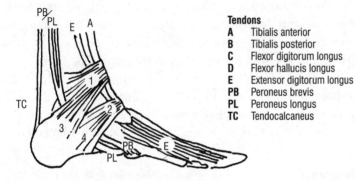

Tendons
A   Tibialis anterior
B   Tibialis posterior
C   Flexor digitorum longus
D   Flexor hallucis longus
E   Extensor digitorum longus
PB Peroneus brevis
PL Peroneus longus
TC Tendocalcaneus

Figure 17.2 Joints and selected ligaments of the foot

The sheer number of joints in the foot means that individual coverage is not always appropriate. In this book the joints and ligaments will be grouped or discussed individually, as required. In this introductory section only a general outline will be given, with the main details appearing under the discussion of 'Movements and muscles' below.

Localized bands of deep fascia constitute the flexor, extensor and peroneal retinacula, which are comparable to the structures of the same name found in the wrist (except for the peroneal retinaculum). Their function is to retain the tendons of the muscles which act over the ankle and foot.

Note: Only the main ligaments on the plantar surface are shown. As in the case of the hand, dorsal ligaments are present, but are to some extent subordinated by the tendons of the extensor muscles. The reader who needs further information will find detailed coverage of the ligaments of all individual joints of the foot, plus details of the retinacula, in Williams (1995).

Movements and muscles

Propulsive movements such as walking are complex, and involve plantar flexion and dorsiflexion at the ankle joint, the movements of inversion and eversion at the subtalar joints and beyond (see below), and flexion and extension at the individual joints of the toes, which are also able to abduct and adduct. The entire foot may also be moved away from the midline in movements that are sometimes termed 'abduction' and 'adduction', but these are really an expression of rotation through the hip and knee joints, with the foot moving like an indicator on a dial.

The movements of the foot inwards (medially) and outwards (laterally) without rotation taking place at hip or knee are known as 'inversion' and 'eversion', and are primarily used in adjustive response to uneven ground. Inversion also incorporates plantar flexion, and eversion incorporates dorsiflexion. The terms 'pronation' and 'supination' are sometimes used rather loosely to imply eversion and inversion, respectively, but the term 'pronation' should really be used strictly for eversion with medial rotation of the knee, and 'supination' for inversion with a laterally rotated knee.

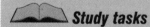

Study tasks

- Highlight the names of the muscles producing inversion and eversion movements.
- Muscle check.
- Read the section on the subtalar joint (below) for details of ligaments that resist excessive inversion/eversion movements.

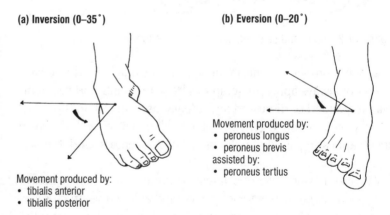

(a) Inversion (0–35°)

Movement produced by:
- tibialis anterior
- tibialis posterior

(b) Eversion (0–20°)

Movement produced by:
- peroneus longus
- peroneus brevis
assisted by:
- peroneus tertius

Figure 17.3 Movements of the foot at the subtalar joints

The subtalar (talocalcanean) joints

The movements of inversion and eversion occur at the joints between the inferior surface of the talus and the superior surface of the calcaneus.

Examination of the inferior surface of the talus reveals a large concave posterior facet which articulates with a reciprocal convex

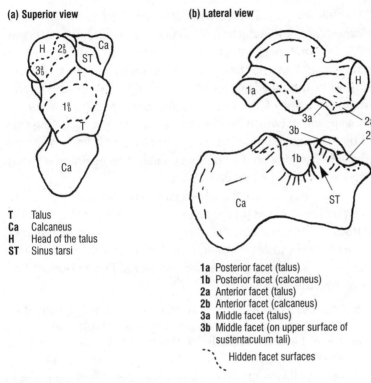

(a) Superior view

(b) Lateral view

T	Talus
Ca	Calcaneus
H	Head of the talus
ST	Sinus tarsi

1a Posterior facet (talus)
1b Posterior facet (calcaneus)
2a Anterior facet (talus)
2b Anterior facet (calcaneus)
3a Middle facet (talus)
3b Middle facet (on upper surface of sustentaculum tali)

Hidden facet surfaces

Figure 17.4 The subtalar joint surfaces

surface straddling the posterior surface of the calcaneus. This is the main portion of the subtalar (talocalcanean) joint.

If the rounded head of the talus is examined it will be noticed that a continuation of the articular surface forms another strip or facet which extends posteromedially towards the large posterior facet described above. A sagittal inspection of this reveals a sharp change in inclination, so that it is possible to divide this surface into an anterior facet (in contact with the head of the talus) and a middle facet which articulates with the sustentaculum tali (ledge) on the calcaneus. There is a deep sulcus in the talus which separates these facets from the posterior facet and which helps to form the sinus tarsi (see below). These surfaces are reciprocated on the calcaneus, and if they are placed together a limited amount of multi-axial gliding is permitted between the talus and calcaneus, which facilitates the movements known as inversion and eversion.

However, the movements of inversion/eversion have an element of rotation, and in order to achieve this the talonavicular articulation is vital. Inspection reveals an almost spheroidal articular surface.

The full functional articulation is referred to as the talocalcaneonavicular joint, which is classified as multi-axial (Williams, 1995). The joint in its entirety consists of the articulations between the medial and anterior facets of the calcaneus and talus, between the head of the talus and the navicular and between the inferior part of the head of the talus and the plantar calcaneonavicular (spring) ligament (which in effect bridges the gap between the talus and the sustentaculum tali of the calcaneus, and is lined by articular hyaline cartilage on its superior surface). This ligament is shown in Figures 17.2(b), 17.7(b) and 16.2(c).

If a bone specimen of the talus is placed on the calcaneus, a cavernous hollow remains on the anterolateral aspect of the joint. This is a result of a deep sulcus that runs between the medial and anterior facets on both the calcaneus and talus and is known as the 'sinus tarsi' (literally 'the tunnel of the foot'). Two important ligaments cross the sinus tarsi:

- The interosseous talocalcanean ligament, which is attached on the inferior surface of the talus and passes inferolaterally to the calcaneus. Since it lies medial to the axis of inversion/eversion it limits eversion.

- The cervical ligament, which passes from the inferolateral part of the neck of the talus to the calcaneus, but lateral to the axis of inversion/eversion. As such, it restricts inversion.

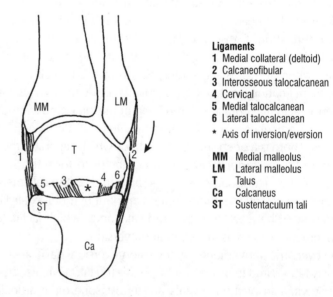

Ligaments
1 Medial collateral (deltoid)
2 Calcaneofibular
3 Interosseous talocalcanean
4 Cervical
5 Medial talocalcanean
6 Lateral talocalcanean

* Axis of inversion/eversion

MM Medial malleolus
LM Lateral malleolus
T Talus
Ca Calcaneus
ST Sustentaculum tali

Figure 17.5 Subtalar joint surfaces (right) (posterior coronal section)

📖 Study tasks

- Shade the ligaments that restrict inversion and eversion, in separate colours.
- With reference to Figure 17.3, consider which muscles will act to restrain excessive inversion and eversion.
- Discover the 'neutral' position of the subtalar joints, which is the position of the foot in which neither inversion nor eversion is taking place (Figure 17.6(a)). This position can be observed from the front by inspecting the curves above and below the lateral malleolus (ankle). These curves should appear equal in the neutral position: deeper or shallower may mean that the foot is still everted or inverted. Practise palpating this position by cupping the heel of a colleague's foot (calcaneus) in one hand and placing your other hand so that your thumb rests over their fourth and fifth toes. The foot can be guided into inversion and eversion, and the neutral position observed.

(a) The neutral position for the subtalar joints

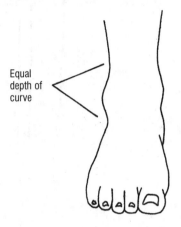

Equal depth of curve

(b) The inversion/eversion gradient at the subtalar joints

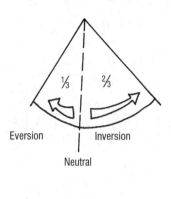

⅓ | ⅔

Eversion | Inversion

Neutral

Figure 17.6 The neutral position of the foot; and inversion/eversion gradients

The transverse tarsal joints

The joint between the head of the talus and the navicular is so closely involved with the anterior part of the subtalar joints that Williams (1995) refers to this as the 'talocalcaneonavicular' joint. In addition, it should be borne in mind that the joint between the calcaneus and cuboid (calcaneocuboid joint) is also functionally part of the inversion/eversion mechanism. Plantar flexion and dorsiflexion of the midfoot also occur through this 'transverse tarsal zone', which has been described by Manter (1941).

The apparent ball-and-socket arrangement between the talus and navicular is in fact limited in scope by the fact that the cuboid and navicular are bound strongly by a fibrous joint or syndesmosis forming the cuboideonavicular joint. The actual movements that take place are dictated by the saddle-shaped calcaneocuboid joint, and this allows a contribution to the movements of inversion/eversion and plantar flexion/dorsiflexion. If these movements are combined, the most likely outcome at the transverse tarsal surfaces are the compound movements of inversion/adduction/plantar-flexion and eversion/abduction/dorsiflexion.

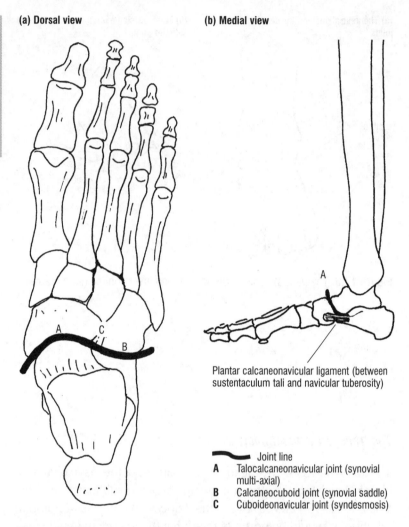

(a) Dorsal view

(b) Medial view

Plantar calcaneonavicular ligament (between sustentaculum tali and navicular tuberosity)

━～ Joint line
A   Talocalcaneonavicular joint (synovial multi-axial)
B   Calcaneocuboid joint (synovial saddle)
C   Cuboideonavicular joint (syndesmosis)

Figure 17.7 The transverse tarsal joints of the foot (after Manter, 1941)

The tarsometatarsal joints

The apparent line of joints between the cuneiform bones, cuboid and metatarsals might also be permitted to include the intertarsal joints.

Inspection of the bones reveals an assembly of wedged shapes coronally, forming a transverse arch (see Figure 17.13). The second metatarsal is recessed between the medial and lateral cuneiforms, making this a stable point for push-off/toe-off in the gait cycle, for standing on tiptoe, and for a ballet dancer standing on points. The second ray of the foot is at once more stable, but less mobile, than the other foot rays. The slight gliding and translation between the individual joint surfaces allows just a few degrees of dorsiflexion and a little more plantar flexion, and accommodates the other movements of the foot described earlier.

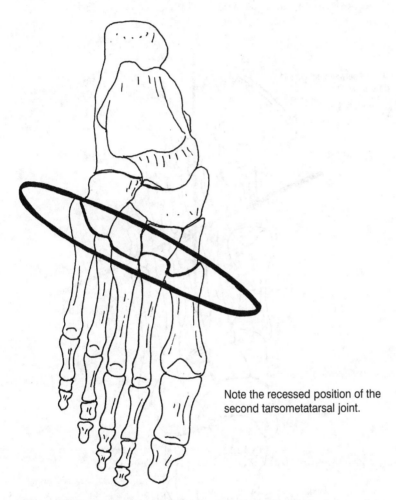

Note the recessed position of the second tarsometatarsal joint.

Figure 17.8 The tarsometatarsal joints of the foot (dorsal view) (synovial approximately plane)

The metatarsophalangeal joints

Individually these are ellipsoid joints that lie proximal to the webs of the toes and as such allow flexion/extension and abduction/adduction (see Figure 17.10). As a group, they help to form a second-class lever that allows push-off/toe off in the gait cycle. The axis that passes through the joints varies considerably between individuals, and forms an angle of 50–70° with the long axis of the foot. It is often referred to as the 'metatarsal break', and represents the generalisation of the instant centres of rotation of the five metatarsophalangeal joints. It has been used conceptually by shoe designers to analyse wear and fit.

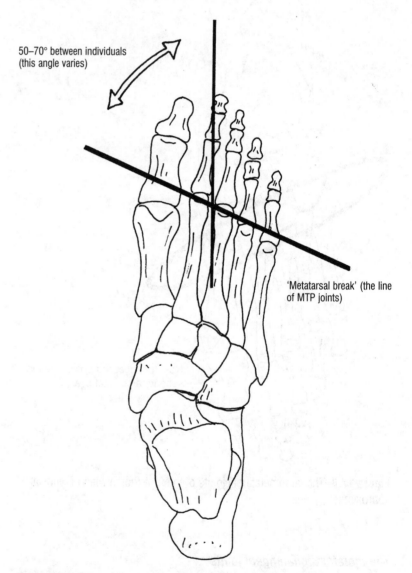

50–70° between individuals
(this angle varies)

'Metatarsal break' (the line of MTP joints)

Figure 17.9 The metatarsophalangeal joints of the foot (dorsal view) (synovial, ellipsoid)

The dominant joint in the metatarsal group is the first metatarsophalangeal joint of the hallux (big toe). The inferior or plantar surface of the head of the first metatarsal has two grooves separated by a ridge in which lie two sesamoid bones. These are embedded in the capsule of the joint incorporating the twin tendons of the flexor hallucis brevis. The movements of the first metatarsophalangeal joint may be taken as representative of the functions of the metatarsophalangeal joints as a group, except that the smaller toes possess a few more degrees of flexion than the hallux. At their limits of 30–40° flexion the joints display a hook-like action which may be used to stabilise the body in precarious positions. They must also extend passively to about 90°, to allow crouching and to accommodate the position of push-off/toe-off in the gait cycle, particularly when it speeds up to a sprint.

Study tasks
- Highlight the names of the muscles producing the movements shown.
- Muscle check.

(a) Flexion
0–40°

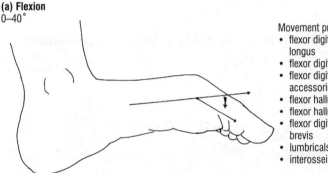

Movement produced by:
- flexor digitorum longus
- flexor digitorum brevis
- flexor digitorum accessorius
- flexor hallucis longus
- flexor hallucis brevis
- flexor digiti minimi brevis
- lumbricals
- interossei

(b) Extension
The normal range of extension is greater than flexion at the metatarsophalangeal joints, especially at the first joint.
0–60° (passively greater)

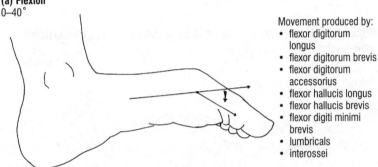

Movement produced by:
- extensor digitorum longus
- extensor digitorum brevis
- extensor hallucis longus

(c) Abduction
Abduction deviates the toes away from the midline represented by the second toe as a reference point.

Adduction returns to reference line

Abduction away from reference line

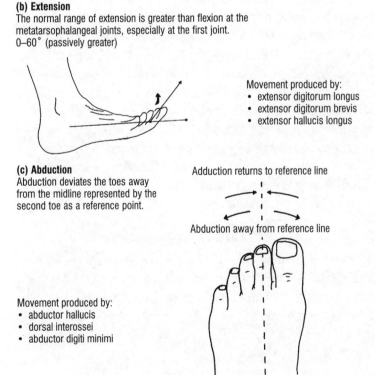

Movement produced by:
- abductor hallucis
- dorsal interossei
- abductor digiti minimi

(d) Adduction
Adduction returns the toes to the midline and is therefore a relative movement from abduction.

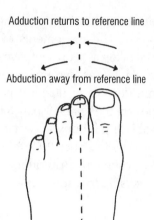

Adduction returns to reference line

Abduction away from reference line

Movement produced by:
• adductor hallucis
• plantar interossei

Figure 17.10 Movements at the metatarsophalangeal and interphalangeal joints of the toes

Clinical Note

The first metatarsophalangeal joint is a common site for the lateral deviation of the big toe seen in the condition known as 'hallux valgus'. For whatever reason (and a diversity of factors ranging from heredity to pointed shoes with raised heels have been implicated), the hallux appears to undergo medial rotation and lateral deviation. Once this process starts, the flexor, extensor and adductor muscles that are attached to the hallux adapt by shortening and pull the hallux farther out of line. At push-off/toe off in the gait cycle the hallux is repetitively pushed into lateral deviation, and the constant friction against footwear encourages the formation of a thick-walled bursa (bunion) and extra bone growth (exostosis) at the medial prominence of the metatarsal head. The lateral collateral ligament shortens and the medial collateral ligament is stretched. Instant centre analysis reveals an altered pattern of motion with sliding replaced by an erratic pattern of distraction and jamming. The condition often accompanies, and may also be caused by, a tendency towards pronation in the foot.

The interphalangeal joints of the toes

The bones that form the toes are similar to those of the hand in that the big toe or hallux has only two phalanges like the thumb, whereas the remaining toes and fingers have three (see Figure 17.2(a)). The pattern of joints and muscle action is therefore similar, though it may be instructive for the reader to observe and consider the many differences and refinements between the weightbearing foot and the normally more precise hand.

The pattern and rhythm of muscular action in the foot are dominated by the needs of weightbearing and locomotion. Flexion of the toes is produced by the powerful flexor digitorum longus and brevis muscles, originating within the calf and foot, respectively. The lumbrical and interossei muscles operate in a similar manner as in the hand. These muscles relax somewhat during the phase in the gait cycle between mid-stance and push-off/toe-off (Figure 16.5), in order to allow passive extension of the toes before the propulsive action of push-off/toe-off when they are reactivated together with the powerful calf plantar flexors.

The other small intrinsic muscles of the toes which produce movements such as abduction and adduction are usually utilised in minor adjustive movements in weightbearing, although the potential for developing skills in the foot similar to those of the hand is often seen in the marvellous adaptability of people suffering from total amelia of the upper limbs.

The ingenuity of the foot lies in its adaptability. It acts both as a powerful supporter of body weight (the static foot) and also as a dynamic lever in movement (the dynamic foot). The adaptability of the foot reflects a number of structural factors:

- the sheer number and variety of bones and joints which do not have a large individual range of movement and are bound quite tightly by ligaments, but collectively allow great strength and movement;
- the way in which changing forces produced by body weight and muscle contraction either stabilise or relax the joints;
- the nature of the arches and plantar fascia;
- the versatility of the muscles.

The special role of the arches

The longitudinal arches

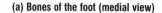

(a) Bones of the foot (medial view)

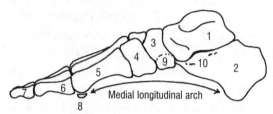

1 Talus
2 Calcaneus
3 Navicular
4 Medial cuneiform
5 First metatarsal
6 Proximal phalanx
7 Distal phalanx
8 Sesamoid bone
9 Tuberosity of navicular
10 Sustentaculum tali (of calcaneus)

(b) Bones of the foot (lateral view)

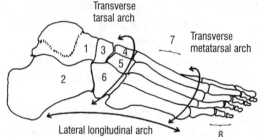

1 Talus
2 Calcaneus
3 Navicular
4 Intermediate cuneiform
5 Lateral cuneiform
6 Cuboid
7 Metatarsals
8 Phalanges

Figure 17.11 The configuration of the arches of the foot

You should notice that eversion flattens the arch, whereas inversion increases it. The reason for this is that the longitudinal arch (especially on the medial side) behaves like a twisted flexible ruler. Test this by pressing one end of a flexible ruler against a hard surface while lifting and twisting the other end. Increasing the amount of twist represents an increase in the medial arch (inversion), whereas a decrease in the twist decreases the arch (eversion). The ability to introduce a controlled increase in the medial arch allows many of the small joints of the foot to be stabilised during propulsion and in turn allows the intrinsic and extrinsic muscles of the foot to produce efficient leverage. The medial arch is almost entirely dependent on this twisted configuration, which is supported by the long and short plantar ligaments, and in the dynamic foot by the action of the tibialis anterior and tibialis posterior muscles. The lateral arch of the foot depends more upon the shape of the bones, and in particular the 'keystone' role of the cuboid, wedged between the calcaneus and the fifth metatarsal (Figure 17.11(b)).

Inversion is accompanied by plantar flexion at the ankle, and if the leg rotates laterally at the knee this produces a supinated foot. In each case, reproduction of these movements on a bone

(a) Medial view

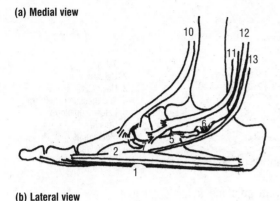

(b) Lateral view

1 Plantar fascia/ aponeurosis
2 Abductor hallucis and intrinsic flexor muscles of foot
3 Abductor digiti minimi
4 Short and long plantar ligaments
5 Plantar calcaneonavicular ligament
6 Interosseous talocalcanean ligament
7 Peroneus longus tendon
8 Peroneus brevis tendon
9 Peroneus tertius tendon
10 Tibialis anterior tendon
11 Tibialis posterior tendon
12 Flexor digitorum longus tendon
13 Flexor hallucis longus tendon

Figure 17.12 The arches of the foot: selected supporting structures

specimen shows that the calcaneus behaves like the twisted end of the flexible ruler. If it is allowed to roll inwards (known as 'valgus') it reduces the twist and flattens the arch. If it becomes more vertical or deviates laterally from the midline (known as 'varus') this increases the arch.

The transverse arches

The proximal transverse tarsal arch across the cuneiform bones and the bases of the metatarsals is really caused by their wedged shape in the frontal plane.

The distal transverse metatarsal arch which runs across the heads of the metatarsal bones mainly reflects the degree of twist in the forefoot, and the dynamics of this arch varies with the degree of pronation and supination which is present. The integrity of this arch is actively aided by the transverse head of the adductor hallucis muscle, which may be strengthened by specific exercises if the transverse arch drops. A chronically pronated 'flat' foot (pes planus) will often show callus (hard skin) formation on the plantar surface under the second and third metatarsal heads.

Bones of the foot (lateral view)

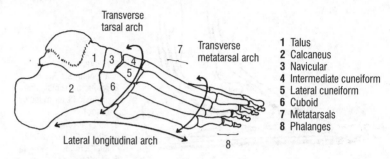

1 Talus
2 Calcaneus
3 Navicular
4 Intermediate cuneiform
5 Lateral cuneiform
6 Cuboid
7 Metatarsals
8 Phalanges

Figure 17.13 The configuration of the arches of the foot

The plantar fascia/aponeurosis

Another structure that increases the stability of the foot and arches during weightbearing and locomotion is the plantar fascia, the deep part of which is often called the 'plantar aponeurosis'. This is composed mainly of collagen, and stretches from the calcaneus as far as the proximal phalanges, serving as an attachment point for the intrinsic foot muscles as well as providing structural support and integrity.

It therefore acts to some extent like a ligament, but being closer to the surface than the short and long plantar ligaments, can exert greater leverage and absorb shock. It has little ability to lengthen, and acts as an effective link between heel and toes, which allows the twisted arch configuration to achieve the necessary degree of stability and rigidity during propulsion.

It achieves all this by combining the qualities of a truss and a windlass (Sammarco, 1989). The aponeurosis, in effect, winds around the heads of the metatarsals, so that when the toes are extended it tightens, and the arch is therefore raised.

(a) Foot flat

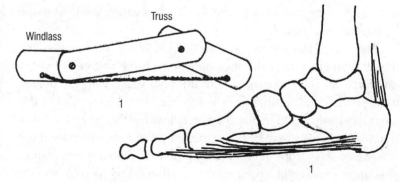

1 The plantar aponeurosis is represented by a cord linking truss and windlass.

(b) Arch raised

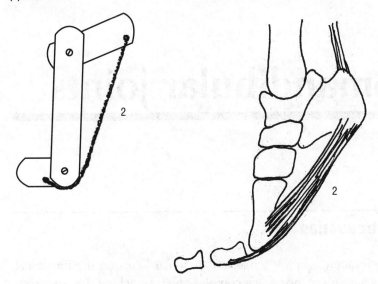

2 Aponeurosis and cord remain same length. Winding up windlass tightens aponeurosis/cord so that the arch is raised by the truss.

Figure 17.14 The truss/windlass mechanism of the plantar aponeurosis

Chapter 18

The temporomandibular joints

Introduction

The influence of the temporomandibular joints on the functional mechanisms of the body is complex and far reaching. The summary given here is brief, and a fuller appreciation of the significance of this area may be obtained by studying the entire stomatognathic system, and also by considering the craniosacral involuntary mechanism. The latter involves the biomechanics of cerebrospinal fluid circulation and its alleged profound effects on the cranium and body as a whole. These subjects are beyond the scope of this book, and only the details of the basic functional anatomy of the temporomandibular joints will be given here. A useful introduction to the temporomandibular joints, with further references, will be found in Hertling (1990). An introduction to the craniosacral involuntary mechanism may be found in Sutherland (1990) and Magoun (1976).

The bones

The mandible, which is the strongest and largest of the facial bones, has a U-shaped body representing two halves that are fused at the midline 'symphysis menti' (already ossified at birth). The two condyles at the heads of the mandible articulate with the mandibular fossae of the temporal bones on each side, thereby forming a right and left temporomandibular joint. The joints always act in concert, and are therefore usually classified as bicondylar, but Williams (1995) favours the term 'ellipsoid'.

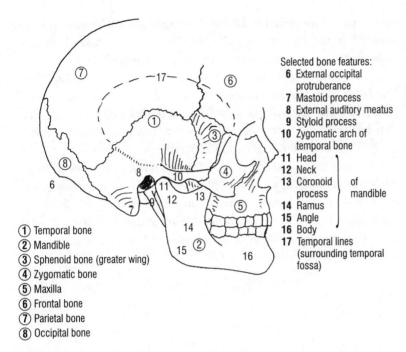

Selected bone features:
6 External occipital protruberance
7 Mastoid process
8 External auditory meatus
9 Styloid process
10 Zygomatic arch of temporal bone
11 Head ⎫
12 Neck ⎪
13 Coronoid process ⎬ of mandible
14 Ramus ⎪
15 Angle ⎭
16 Body
17 Temporal lines (surrounding temporal fossa)

① Temporal bone
② Mandible
③ Sphenoid bone (greater wing)
④ Zygomatic bone
⑤ Maxilla
⑥ Frontal bone
⑦ Parietal bone
⑧ Occipital bone

Study tasks

• Shade the mandible and temporal bone in separate colours and note the features shown.

Figure 18.1 The temporomandibular joint: surrounding bones and bone features (lateral view)

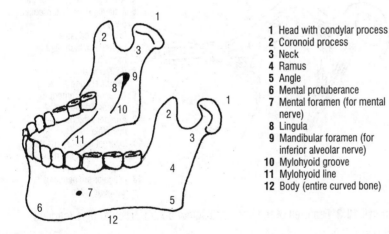

1 Head with condylar process
2 Coronoid process
3 Neck
4 Ramus
5 Angle
6 Mental protuberance
7 Mental foramen (for mental nerve)
8 Lingula
9 Mandibular foramen (for inferior alveolar nerve)
10 Mylohyoid groove
11 Mylohyoid line
12 Body (entire curved bone)

Study tasks

• Highlight the bony features of the mandible as shown, and obtain a bone specimen or observe the features on an articulated skeleton.
• Palpate on yourself as many of these features as you can.

Figure 18.2 The mandible and selected features (oblique lateral view)

The joints and ligaments

The temporal fossae that accept the mandibular condyles are deep, concave and oval transversely. This allows them to receive the condyles when the jaw is closed, and also to facilitate the side-to-side or lateral movements involved in mastication (chewing). However, when the jaw opens, the condyles translate forwards over a convex surface known as the 'articular eminence', and an intra-articular fibrocartilage disc, or meniscus, is present to facilitate this. The latter is, in reality, a complex structure, concave on its inferior surface for the mandibular condyles, and concavoconvex (saddle shaped) on its superior surface for the temporal articulation. It is not at all like the other menisci and fibrocartilaginous discs in the body, and fits over the mandibular condyles rather like a cap. The joint on each side is effectively divided into two parts by this structure, and the mandible and temporal fossae are not directly in contact with each other but rather with the opposing surfaces of the articular disc. The articular surfaces of the mandible and temporal fossa are covered by white fibrocartilage.

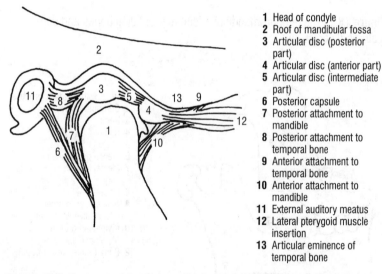

1 Head of condyle
2 Roof of mandibular fossa
3 Articular disc (posterior part)
4 Articular disc (anterior part)
5 Articular disc (intermediate part)
6 Posterior capsule
7 Posterior attachment to mandible
8 Posterior attachment to temporal bone
9 Anterior attachment to temporal bone
10 Anterior attachment to mandible
11 External auditory meatus
12 Lateral pterygoid muscle insertion
13 Articular eminence of temporal bone

Figure 18.3 Features of the fibrocartilaginous intra-articular disc (sagittal view)

The fibrous capsule surrounds the joint, and is lax enough to allow for movement, but is taut inferiorly between the mandibular head and meniscus. This has the effect of drawing the meniscus forwards when the mandible is lowered or depressed.

(a) Lateral view (right)

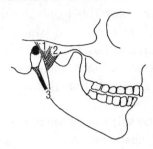

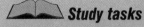

Study tasks

- Shade the ligaments in colour, and highlight their names.
- Consider their function.

1 Capsule
2 Lateral ligament
3 Stylomandibular ligament
4 Sphenomandibular ligament
5 Mylohyoid groove
6 Lingula
7 Mylohyoid line (part)

(b) Medial view (left)

Figure 18.4 Ligaments of the temporomandibular joint

Movements and muscles

The basic movements of the temporomandibular joints are discussed below, but the actual dynamics of the movements are complex and may need further elaboration.

The 'anatomical position' implies a neutral position of rest for the temporomandibular joints. This requires the head to be upright and the muscles acting over the joints to be in a state of equilibrium. The teeth should not be in contact but should express a slight 'interocclusal clearance' (a gap of 2–5 mm at the incisors). The tongue should be in slight contact against the palate with the tip resting against the anterior palate, almost touching the central incisors. All associated tissues and muscles of the stomatognathic system should be in a state of rest.

Any disturbance of this state of equilibrium (for example, teeth in contact, clenched or grinding) when the system should be at rest implies a waste of energy which will contribute to fatigue, poor repair/healing and an increased rate of degeneration in the tissues involved.

Occlusion (closure) of the jaw is the end point of elevation of the mandible, and may involve considerable forces of contraction in

Study tasks

- Observe a colleague's relaxed jaw, or your own in front of a mirror. If the lips are parted the resting position may be observed. The interocclusal gap may be tested by gently lifting the chin to produce occlusion.
- Gently elevate (close) the jaw and look for proper occlusion with intercuspation as described above.
- An appreciation of movement at these joints may be obtained by gently placing your finger tips in your ears (external auditory meatus) while moving your jaw as described below. Clicks or crepitus indicate malfunction of the joint, usually involving the disc.

 Study tasks

- Highlight the names of the muscles shown in Figure 18.5.
- Muscle check.
- Review the terms 'synergist' and 'fixator' as applied to muscles.
- Consider which tissues restrain excessive movement in this case.

the relevant muscles (for example, the masseter and temporalis) if powerful mastication (chewing) is required. The mandible and maxilla should be in a balanced midline position in which the upper and lower teeth meet in a comfortable 'best fit' manner, with the lower incisors resting inside and posterior to the upper incisors (a position known as 'intercuspation'). Departure from this 'best fit' position is termed 'malocclusion'.

The elasticity of the menisci, and the structure of the temporomandibular joints and their suspensory ligaments allow a considerable freedom of movement, albeit within limits imposed by the ligaments.

From the starting position of rest, the mandible may be moved in a sagittal, frontal and horizontal plane or a combination of these. This allows the movements of elevation and depression, protrusion (protraction) and retraction (retrusion), lateral movements, and a composite of all these which is the generalised movement of mastication. As isolated movements, some of these appear strange, but it should be appreciated that they are normally utilised as composite movements in activities such as eating. Each of these movements of the mandible will now be examined in turn:

Depression

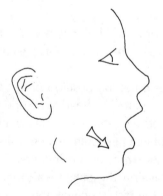

Movement produced by:
- lateral pterygoid

assisted by:
- gravity
- digastric
- mylohyoid
- geniohyoid

Figure 18.5 Depression of the mandible

The mouth opens initially under the influence of gravity and with the aid of the suprahyoid muscles. As the mandibular condyles rotate around a transverse axis, the action of the lateral pterygoid muscles on each side pulls the condyles and the attached meniscus forwards on to the articular eminence of the temporal bone. Translation continues anteriorly until stretch receptors in the capsule, ligaments and meniscus signal discomfort.

When the lateral pterygoids draw the mandible forwards, a downward and backward component is provided by the contraction of the suprahyoid muscles (those attached to the underside of the mandible and to the upper surface of the hyoid bone, which lies below the mandible). These synergistic muscles are the digastric, mylohyoid, geniohyoid and stylohyoid. The infrahyoid muscles (those attached below the hyoid to its inferior surface) act as fixators holding the hyoid platform steady while the suprahyoid and lateral pterygoid muscles act directly on the mandible. The infrahyoid muscles which act as fixators in this process are the thyrohyoid, omohyoid, sternohyoid and sternothyroid.

Elevation

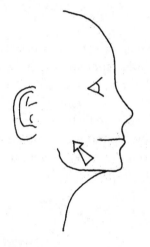

Movement produced by:
- masseter
- medial pterygoid
- temporalis

Figure 18.6 Elevation of the mandible

Closure of the mouth implies a reversal of the process of depression of the mandible. The deeper posterior fibres of the masseter, and posterior fibres of temporalis muscles initiate backward gliding of the mandibular condyles. The condyles are able to hinge on the menisci in the early stages of closure, since these are still held forwards by the gradually relaxing lateral pterygoids.

The process continues with the increasing activity of the main elevating muscles, the masseter, medial pterygoid, and temporalis, which together provide an extremely powerful bite when necessary, resulting in intercuspation and full occlusion of the upper and lower teeth.

Protrusion (protraction)

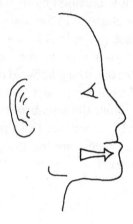

Movement produced by:
- lateral pterygoid
- medial pterygoid
- masseter (superficial fibres)

Figure 18.7 Protrusion of the mandible

From the position of rest, the mandible translates forwards so that the condyles and menisci reach the articular eminence of the temporal bone, but no rotation around a transverse axis takes place, which differentiates this movement from depression of the mandible. This forward movement is effected by the contraction of the lateral pterygoids, with antigravity support (eccentric contraction) from the medial pterygoids and masseter to prevent the jaw from dropping.

Retraction (retrusion)

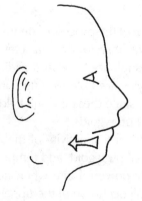

Movement produced by:
- temporalis (posterior fibres)

assisted by:
- masseter (middle and deep fibres)
- geniohyoid
- digastric

Figure 18.8 Retraction of the mandible

This is the return movement to the position of rest from protrusion, though the movement may be continued further as a signal of facial expression. It is protrusion in reverse, with the posterior fibres of temporalis drawing the mandible back, assisted by the deep fibres of the masseter and suprahyoid muscles.

Lateral movements

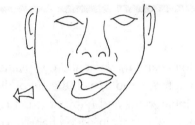 

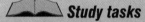

Movement produced by:
• medial pterygoid ⎫
• lateral pterygoid ⎭ one one side, acting alternately

assisted by:
• temporalis (posterior fibres) – on the opposite side
• masseter

Figure 18.9 Lateral movements of the mandible

These are asymmetrical side-to-side movements utilised in mastication. On one side the mandibular condyle and meniscus translate forwards and downwards pulled by the lateral and medial pterygoid muscles (as in protrusion on one side only), with some medial deviation. The other condyle remains in the temporal fossa, but rotates in a lateral direction with some medial translation caused by the contraction of the posterior fibres of temporalis. These complex movements are probably also balanced and guided by the other muscles of mastication, such as the masseter, as well as the synergistic muscles of the suprahyoid group.

Mastication (chewing)

This is a composite movement involving proprioception and acquired skills that coordinate all movements of the mandible: that is, depression and elevation, with protrusion and retraction, plus lateral movements. Muscles of the tongue and cheek (for example, the buccinator) are also utilised to keep food moving over the teeth.

Glossary of terms and abbreviations

Many of the terms relevant to joints and movement are explained in the introductory chapters (Chapters 2 and 3) with the aid of diagrams. The names of the individual joints and relevant structures usually appear in the introductory diagrams in each chapter. The index may also be consulted.

A

Abduction   Movement of the body parts away from the midline.

Accessory process   A small bony prominence at the root of the transverse process of a lumbar vertebra.

Active   As in movements carried out using muscles under voluntary control.

Adduction   Return movement of the body parts towards the midline.

Adjunct   As in a movement (usually rotation) in a joint which is produced by muscular action, gravity or external forces.

Afferent   Conveying towards a centre.

Amelia   Congenital absence of one or more limbs.

Anastomosis (anastomosing)   Intercommunication of the branches of two or more arteries/veins.

Anatomical position   The universal position of the body accepted as a starting point of reference for anatomical description. It assumes that the body is standing facing the observer with palms facing anteriorly, legs together and arms at the sides. All movements are described from this position.

Anatomical snuffbox   The small hollow formed by the extensor and abductor tendons of the thumb in the extended hand. Snuff (tobacco) can be taken from this point. It is not recommended.

Angle   Change in direction of bone, as in the ribs, mandible, scapula.

Antagonistic   A term applied to muscles that produce the opposite movement to a specified movement taking place.

Anteversion   An anterior twist or tilt, seen, for example, in the angle of torsion of a bone.

Aponeurosis   A flat sheet of tendon connecting one muscle to another, or to bone.

Apophyseal/Zygapophyseal   The offshoot joints of the articular processes (facets) of vertebrae, which help to define the posterior part of a motion segment.

Appendicular   The peripheral parts of the skeleton, including the limbs and limb girdles.

Arc   A curved line which does not represent the shortest path between two points.

Arthrokinematics   The branch of mechanics that deals with the motion of joints without reference to the motion of the bones.

Articulation   A joint, or movement between two or more bones.

Asymmetry   The absence of symmetry.

Atypical   Not typical, as in 'atypical synovial joint'.

Avascular   Without blood supply.

Axial   The central parts of the skeleton, including the vertebral column, cranial bones, sternum and ribs.

Axis   An imaginary line about which the body or a body part moves or rotates.

B

Belly   With reference to muscle, it means the fleshy part of a muscle.

Bi-   Two or both.

Bi-axial   Two axes of movement.

Bifurcate   To split into two parts.

Bilateral (bilaterally)   On both sides.

Biomechanics   The study of mechanical motion in an organism.

Bipedal   Pertaining to two feet; two-footed.

Bursa   A pouch of synovial membrane, usually located near joint surfaces in order to minimise friction.

Bursitis   Inflammation of a bursa.

C

c.   Circa (around).

C1–7   Denotes cervical vertebrae, of which there are seven. The numbers may refer to either the root value of the cervical nerves (of which there are eight) or to individual vertebral segments, depending on the context.

Capsule   The ligamentous fibres forming a casing around synovial joints. It is lined by synovial membrane.

Capsulitis   Inflammation of a capsule.

Cardiovascular   Relating to heart and circulation.

Carpus (Carpal)   The collective term for the wrist.

Cartilage   The connective tissue that covers the articular surfaces of bones, or joins bones together. There are two types: hyaline and fibrocartilage, and they also provide the constituent material for certain body parts, such as the auricle (pinna) of the ear, intervertebral discs and menisci.

Cartilaginous joint   Two articulating bones connected by either hyaline cartilage (a primary cartilaginous joint or synchondrosis) or by fibrocartilage (a secondary cartilaginous joint or symphysis).

Centre of gravity   A point at the exact centre of an object's mass. An imaginary line may be drawn through the body to link these points.

Cervical   Relating to the neck.

Chord   A line that follows the shortest possible path between two points.

Circumduction   A movement of the body or a part of the body, in which the distal part describes the arc of a circle while the proximal attachment remains stable. It combines elements of extension, abduction, flexion and adduction.

Close-packed   The position in which joint surfaces are fully congruent (fit best) with optimal support from capsule and ligaments.

Collagen   A white protein which is the main organic constituent giving strength to connective tissue.

Collateral   At the sides.

Compression   A mode of loading in which equal and opposite loads are applied towards the surfaces of a joint.

Compound   As applied to joints, possessing more than one pair of articulating surfaces.

Concave   A hollow shape.

Condyle   A large articular prominence on a bone.

Condylar (bicondylar) joint   Uni-axial synovial joint that nevertheless permits a degree of rotation and is formed between the rounded convex condyle(s) of one articular bone and the concave accepting surface of the other.

Congenital   Existing at or before birth.

Conjunct   As in an automatic movement (usually rotation) that necessarily accompanies a particular joint movement because of the shape of the articular surfaces.

Congruence (congruity)   Mutual fit between two surfaces.

Connective tissue   Abundant tissue that supports, binds and gives physical integrity to the structures of the body.

Contraction   With reference to muscle, the shortening action of the sliding filaments which produces movement.

Contralateral   On the opposite side.

Convergent   Approaching a focal point from a more dispersed (divergent) starting point.

Convex   A rounded shape.

Costal   Relating to the ribs.

Cranium (cranial)   The skull; relating to the skull.

Creep   The progressive deformation of matter owing to constant low loading over an extended period of time.

Crepitus   A crackling sound or sensation.

Crest   A prominent ridge or border on a bone.

Cruciate   Crossing or cross-like.

D

Deep   Below a more superficial structure

Degrees of freedom   The number of different ways in which a joint can move.

Depression   Movement that lowers a part of the body. Hollow.

Dislocation   The displacement of a part, usually the bony part of a joint.

Distal   Further from a point of attachment which is closer to the midline of the body.

Distraction   The specific movement of two joint surfaces away from each other.

Dorsal expansion   A small aponeurosis covering the dorsal surface of the proximal phalanges of both the hands and feet, and providing important muscle attachments.

Dorsiflexion   Flexion of the foot at the ankle joint in an upward direction.

Dorsum (dorsal)   The back or posterior surface.

Dura mater   The outer membrane covering the brain and spinal cord.

Dynamics   The study of forces that act on a body in motion.

E

Eccentric   In the context of muscle contraction, an increase in the distance between origin and insertion, due to the graded relaxation of some muscle fibres.

Efferent   Conveying away from a centre.

Elastin   A flexible yellow protein which gives elasticity to connective tissue.

Elevation   Movement that raises part of the body.

Ellipsoid joint   A bi-axial synovial joint that has oval-shaped articular surfaces allowing flexion/extension and abduction/adduction.

Elliptical   Shaped like a regular oval.

Epi-   Above or outer.

Epicondyle   A prominence above the condyle on a bone.

Epicondylitis   Inflammation at the site of an epicondyle.

Epimysium   The protective cover surrounding the belly of a muscle.

Eversion   The outward movement of the sole of the foot.

Extension   An increase in the anterior angle between articulating surfaces, which brings posterior surfaces closer together (except in the case of the knee, ankle and toes).

Extensor expansion   See 'Dorsal expansion' above.

Extrinsic   Coming from or originating from the outside.

F

Facet   A smooth, flat, well-demarcated surface on a bone.

Fascia   Strong fibrous tissue that forms compartments between muscle. Its function is still not fully understood, but it should not be regarded as inert or totally non-contractile.

Fibrocartilage   The fibrous cartilage that is the main constituent of certain body parts, such as menisci and intervertebral discs. In some joints it covers the articular surfaces between bones instead of hyaline cartilage, but this is not typical.

Flexion   The folding movement of a joint which brings the anterior surfaces closer together and decreases the anterior angle between the bones (except in the case of the knee, ankle and toes).

Foramen (foramina)   An opening through which neurovascular structures may pass.

Force   A physical quantity that may accelerate/deform a body.

Fossa   A depression on a bone or tissue surface.

G

Gliding   The adjustive sliding movement between synovial plane joints, such as the intercarpal and tarsal bones of the wrist and foot.

Glycosaminoglycan   Long chains of repeating disaccharide units found in proteoglycans.

H

Head   The term is applied to muscle and bone. With reference to muscle, it means one of two or more proximal attachments to a bone. With reference to bone, it is the rounded articular part at one end, supported by a narrower shaft.

Hinge joint   A uni-axial synovial joint where the concave/convex articulating surfaces permit only flexion/extension.

Hyaline cartilage   A pliable type of cartilage, almost translucent in appearance, which covers the articular surfaces of most synovial joints, and also forms costal (rib) cartilage.

Hydrostatic   The pressure on liquid at rest due to the weight of the liquid surrounding it. Liquid in a state of equilibrium.

Hyper-   Above.

Hypertonia   Excessive muscle tone.

Hypertrophy   The enlargement of tissue due to an increase in the size of its constituent cells.

Hypo-   Below.

I

Inferior   Towards the lower part of a structure; in a downward direction.

Infra-   Beneath.

Inguinal   Relating to the groin area.

Insertion   With reference to muscles, the attachment point that moves the most, and is drawn towards the less mobile attachment point known as the origin, when a muscle contracts.

Inter-   Between

Interosseous   Between bones.

Intra-   Within.

Intrinsic   From the inside.

Inversion   The inward movement of the sole of the foot.

In vivo   Within the living body.
Ipsilateral   On the same side.

K

Kinematics   The branch of mechanics that deals with the motion of a body without reference to force or mass.
Kinesiology   The study of motion in the human body.
Kyphosis   The convex flexion curve found in the thoracic spine, as viewed from the side.

L

L1–5   Denotes lumbar vertebrae, of which there are five. The numbers may refer either to the root value of the lumbar nerves or to individual vertebral segments, depending on the context.
L/S   The lumbosacral disc area, which may also be designated L5–S1.
Labrum   Rim.
Lamella   (lamellae) Thin leaf or plate.
Lateral   Farther away from the midline of the body or a structure.
Ligament   Dense fibrous tissue that connects bone to bone.
Line   With reference to a bone, a slightly raised ridge.
Linea aspera   'Roughened line'(Latin translation).
Longitudinal   Lying lengthways; it often refers to an axis of movement which runs along the shaft of a long bone, or from superior to inferior through the body.
Loose-packed   A position in which joint surfaces are not congruent and the capsule and ligaments are lax.
Lordosis   The concave extension curve found in the cervical and lumber spine, as viewed from the side.
Lumbar   Relating to the low back, and specifically the vertebral column at the L1–5 level.

M

Mamillary process   A rounded rough prominence found on the posterior border of the superior articular processes of lumbar vertebrae.
Mastication   Chewing.
Mastoid process   A breast-shaped process on the temporal bone, behind the ear.
Medial   Closer to the midline of the body or a structure.

Median plane The imaginary flat surface that runs through the midline or middle, and divides the body into two equal left and right halves.

Mediastinum Partition between the lungs.

Meniscus (menisci) Crescent- or moon-shaped fibrocartilage disc(s) found within synovial joints.

Meta- Denotes a state or area of change.

Metabolic The biochemical changes that occur in the body.

Modus operandi The way in which something works.

Mortise (and tenon) A carpentry term to describe a piece of wood which has been precisely cut or drilled to accept another piece.

Motion segment A functional unit of the spine, defined as two adjacent vertebrae with their connecting ligaments.

Multi- Many.

Multi- (poly)-axial More than two axes of movement.

Myo- Referring to muscle.

N

Neck With reference to a bone, the narrow part supporting the head.

Necrosis Morphological changes induced by cell death.

Neuro- Pertaining to nerves and the nervous system.

Neuropathy A disease or adverse condition affecting the nervous system.

Newton (N) The International System (SI) unit of force which if applied to a mass of 1 kg gives it an acceleration of 1 m/sec^2.

Nuchae (ligamentum nuchae) Of the nape of the neck.

O

Orthotic A device that improves posture or function by supporting a body structure (for example, the foot).

Osteoarthritis A degenerative joint condition featuring the erosion of articular cartilage with changes in the nature of articular bone surfaces and other features.

Oblique At an angle.

Obstetrics The management of pregnancy, labour, and the puerperium.

Occlusion The act of shutting or closing.

Opposition Flexion of the thumb towards the flexed fingertips, particularly between the thumb and little finger.

Origin With reference to muscles, the more stable attachment point that draws the more mobile insertion towards itself when a muscle contracts.

Osteochondrosis A disease of various ossification centres in immature bones, which leads to deformation.

Osteopathic Pertaining to osteopathy, a patient-centred system of medicine founded by an American, Andrew Taylor Still, based on the theory that the health of mind and body depends on normal balanced mental and physical structure, favourable environment, and nutrition. Osteopaths aim to restore health by a process of diagnosis using palpation, appropriate diagnostic examination, and treatment which may involve manipulation of body structures and advice about lifestyle and environment.

Osteoporosis Reduction in bone mass which leads to weakening in the structure, often resulting in fracture after relatively minor trauma,

P

Palmar Referring to the palm or anterior surface of the hand.

Palpation (palpate) Feel or gently touch for the purpose of evaluation or diagnosis.

Parturition The expulsive act of giving birth.

Passive As in movements of a body part which are not performed by the active use of the muscles that control that body part.

Phalanx (phalanges) Small bone(s) of the fingers and toes.

Pivot joint A uni-axial synovial joint that permits only rotatory types of movement.

Plane An imaginary flat surface that divides the body into sections.

Plane joint A synovial joint in which the surfaces are either flat or slightly curved, permitting only sliding or gliding movements.

Plantar The undersurface of the foot.

Plantar flexion The downward movement of the foot at the ankle joint which points the toes.

Plexus Network (of blood vessels or nerves).

Posterior Towards or at the back of the body.

Prehensile Adapted to gripping and grasping.

Process A prominent projection on a bone.

Pronation The rotatory movement of the forearm which allows the palm of the hand to face posteriorly. In the foot the equivalent is eversion, but this is not produced by rotation.

Prone Lying horizontally, face downwards.

Proprioception   The reception of information from nerve endings concerning movement and positioning of the body.

Proteoglycan   A protein–carbohydrate complex with a protein core to which glycosaminoglycan (disaccharide) chains are bound.

Protraction   The movement of the scapula (shoulder blade) or mandible (lower jaw) in an anterior direction.

Proximal   Nearer to a point of attachment which is closer to the midline of the body.

Q

Quadrate   Square.

Quadrilateral   Four-sided, but with only two parallel sides.

Quadrupedal   Pertaining to four feet; four-footed.

R

Radiograph/Roentgenograph (-gram)   An image produced on a film by the passage of X-rays, and popularly known as an 'X-ray.'

Ramus (rami)   Root(s)

Raphe   With reference to muscle, a tendinous seam between muscle tissue.

Retinaculum   A structure that retains tissue or an organ in place.

Retraction   The movement of the scapula (shoulder blade) or mandible (lower jaw) in a posterior direction.

Retro-   Lying behind.

Retroversion   A posterior twist or tilt, seen for example in the angle of torsion of a bone.

Rhombus (rhomboid)   Four equal sides, parallel but slanted, unlike a square.

Roll   The movement of a curved articular surface over another in a manner similar to the rolling motion of a wheel.

Rotation   Movement around a longitudinal axis in either an inward direction (medial/internal rotation) or an outward direction (lateral/external rotation); right or left in the vertebral column.

S

S1–5   Denotes the sacrum, which has five fused segments. The numbers may refer to either the root values of the sacral nerves or to the individual sacral segments, depending on the context.

Saddle (sellar) joint   A bi-axial synovial joint in which the articular surfaces are concave in one direction and convex in the other, as seen on a saddle. This allows flexion/extension and abduction/adduction movements and a very limited amount of rotation.

Scoliosis   A significant lateral deviation in the normally straight vertebral column.

Septum   A dividing wall or partition.

Sesamoid   A relatively small bone found within a tendon (the name means 'seed-like'). Its purpose is to increases leverage, or alter the character of muscular pull.

Shaft   The narrow tubular part of a long bone.

Shear   A loading mode applied parallel to the surface of a structure (joint) which causes angular deformation.

Slide (glide)   The movement of a curved articular surface over another in a manner similar to a skidding wheel. In the case of flatter articular surfaces the term is synonymous with translation.

SP (SPs)   An abbreviation denoting the spinous process(es) of individual vertebrae.

Spin   The movement of a bone around its stationary mechanical axis.

Spine   A sharp slender projection of bone.

Spheroidal joint (ball-and-socket joint)   A multi-axial synovial joint between the ball-shaped head of a bone and the socket-shaped surface of another. It allows all ranges of movement.

Stomatognathic   Pertaining to the mouth and jaws.

Striated   Striped.

Subcutaneous   Just beneath the skin.

Subluxation   An incomplete/partial dislocation.

Sulcus   With reference to bones, a groove or depression.

Superficial   At or relatively near the surface.

Superior   Towards the upper part of a structure; in an upward direction.

Supination   The rotatory movement of the forearm which allows the palm of the hand to face anteriorly. In the foot the equivalent is inversion, but this is not produced by rotation.

Supine   Lying horizontally, face upwards.

Supra-   Above.

Swing   A movement of a bone involving a shift in its mechanical axis.

Symmetry   The similar or equal relationship of parts on each side of a plane of the body or around a common axis.

Symphysis   A secondary cartilaginous joint in which the articulating bones are held together by fibrocartilage.

Syndesmosis   A joint in which the articular surfaces are held together by strong fibrous tissue, allowing only strictly limited movement.

Syndrome   A recognised set of symptoms that occur together producing a response typical of a particular health problem.

Synergist   An assisting muscle that has a stabilising effect and reduces undesirable movement.

Synovium (synovial)   The membrane that lines joint capsules and secretes a lubricating fluid.

T

T1–12   Denotes thoracic vertebrae, of which there are 12. The numbers may refer to either the root value of the thoracic nerves or to individual vertebral segments, depending on the context.

Tarsus (tarsal)   A collective term for the posterior part of the foot.

Tendon   Dense white connective tissue that joins muscle to bone.

Tension (tensile)   Loading that produces equal and opposite forces away from the surface of a structure, resulting in lengthening.

Thenar   Refers to the palm of the hand.

Thoracolumbar   A junction area between the thoracic and lumbar regions.

Thorax (thoracic)   The chest; 'thoracic' may refer to the middle part of the vertebral column from T1 to T12 where the ribs attach.

Tonic   With reference to muscle, this describes the normal resting state.

Torsion   Loading that causes a structure to twist about an axis, subjecting it to a combination of shear, tension and compression.

Torso   The trunk, or that part of the body excluding head and limbs.

Torque   The force that produces torsion.

TP (TPs)   An abbreviation denoting the transverse process(es) of individual vertebrae.

Trabeculae   A lattice of thin lines showing patterns of force in spongy bone.

Traction   A drawing or pulling force, often applied along the axis of a structure.

Translation   The parallel motion (slide/glide) of one articulating surface over another.

Transverse   Across, or at right angles to the length of a structure.

Trapez-   Suggests a four-sided figure with only one pair of parallel sides (from 'trapezium').

Trochanter   A large blunt projection found near the head of the femur (thigh bone).

Tubercle   A small rounded projection on a bone.

Tuberosity   A large rounded, often roughened projection on a bone.

U

Unciform/uncinate   A hook-shaped lip on the superior and lateral border of a cervical vertebral body.

Uni-   One or single.

Uni-axial   One axis of movement only.

Unilateral   On one side only.

V

Valgus   Displacement or angulation away from the midline.

Varicose   Unnaturally and permanently enlarged.

Varus   Displacement or angulation towards the midline.

Vascular   Pertaining to blood vessels.

Vasomotor   Affecting the calibre of a blood vessel.

Volar   On a palmar or plantar surface.

W

Whiplash (injury)   An acceleration/deceleration injury to the upper cervical spine caused by the lower cervical spine and vertebrae below acting like the handle of a whip while the head and upper cervical spine behave in the manner of a mobile lash.

References

Adams, M.A. (1994) Biomechanics of the lumbar motion segment. In: *Grieve's Modern Manual Therapy: The Vertebral Column*, 2nd edn. (ed. Boyling, J.D. and Palastanga, N.) Churchill Livingstone, Edinburgh.

Adams, M.A. and Hutton, W.C. (1981) The relevance of torsion to the mechanical derangement of the lumbar spine. *Spine*, **6**, 241–248.

Adams, M.A. and Hutton, W.C. (1982) Prolapsed intervertebral disc. A hyperflexion injury. *Spine*, **7**, 184–191.

Adams, M.A. and Hutton, W.C. (1983) The effect of posture on the fluid content of lumbar intervertebral discs. *Spine*, **8**, 665–671.

Adams, M.A., Dolan, P. and Hutton, W.C. (1987) Diurnal variations in the stresses on the lumbar spine. *Spine*, **12**, 130–137.

Barnett, C.H. *et al.* (1961) *Synovial Joints, Their Structure and Mechanics*. Longman, London.

Basmajian, J.V. (1969) Recent advances in the functional anatomy of the upper limb. *American Journal of Physical Medicine*, **48**, 165–177.

Basmajian, J.V. and Latif, M.A. (1957) Integrated actions and functions of the chief flexors of the elbow. *Journal of Bone Joint Surgery*, **39A**, 1106–1118.

Bourdillon, J.F., Day, E.A. and Bookhout, M.R. (1992) *Spinal Manipulation*, 5th edn. Butterworth Heinemann, London.

Cailliet, R. (1981) *Low Back Pain Syndrome*, 3rd edn. F.A. Davis, Philadelphia.

Cailliet, R. (1991a) *Neck and Arm Pain*, 3rd edn. F.A. Davis, Philadelphia.

Cailliet, R. (1991b) *Shoulder Pain*, 3rd edn. F.A. Davis, Philadelphia.

Cailliet, R. (1995) *Low Back Pain Syndrome*, 5th edn. F.A. Davis, Philadelphia.

Čihák, R. (1970) Variations of lumbosacral joints and their morphogenesis. *Acta Univ. Carolin. Med.*, **16**, 145–165.

Codman, E.A. (1934) *The Shoulder*. Thomas Todd, Boston.

Cossette, J.W., Farfan, H.F., Robertson, G.H. and Wells, R.V. (1971) The instantaneous centre of rotation of the third lumbar intervertebral joint. *Journal of Biomechanics*, **4**, 149–153.

Currier, D.P. (1972), Maximal isometric tension of the elbow extensors at varied positions. Part 2. Assessment of extensor components by quantitative electromyography. *Physical Therapy*, **52**, 1265–1276.

De Pukey, P. (1935) The physiological oscillation of the length of the body. *Acta Orthop. Scand.*, **6**, 338.

Dolan, P., Adams, M.A. and Hutton, W.C. (1987) The short term effects of chymopapain on intervertebral discs. *Journal of Bone and Joint Surgery*, **69B**, 422–428.

Downing, C.H. (1923) *Principles and Practice of Osteopathy*. Williams, Kansas City.

Frankel, V.H. and Nordin, M. (1989) Biomechanics of the ankle. In: *Basic Biomechanics of the Musculoskeletal System*, (ed. Nordin, M. and Frankel, V.H.). 2nd edn. Lee & Febiger, Philadelphia, 153–161.

Fryette, H.H. (1954) *Principles of Osteopathic Technic*. Academy of Applied Osteopathy, Carmel, CA.

General Osteopathic Council (1999) *Standard 2000* (S2K), GOsC, London.

Giles, L. and Taylor, J.R. (1987) Human Zygapophyseal Joint Capsule and Synovial Fold Innervation. *British Journal of Rheumatology*, **26**, 93–98.

Gunzburg, R., Hutton, W.C. and Fraser, R. (1991) Axial rotation of the lumbar spine and the effect of flexion. *Spine*, **16**, 22–29.

Hartman, L. (1997) *Handbook of Osteopathic Technique*, 3rd edn. Stanley Thornes, Cheltenham.

Hazelton, F.T., Smidt, G.L., Flatt, A.E. and Stephens, R.L. (1975) The influence of wrist position on the force produced by the finger flexors. *Journal of Biomechanics*, **8**, 301.

Hertling, D. (1990) The temporomandibular joint. In: *Management of Common Musculoskeletal Disorders*, 2nd edn. (ed. Hertling, D. and Kessler, R.M.). J.B. Lippincott, Philadelphia.

Hertling, D. and Kessler, R.M. (1990) The wrist and hand complex. In: *Management of Common Musculoskeletal Disorders*, 2nd edn. (ed. Hertling, D. and Kessler, R.M.). J.B. Lippincott, Philadelphia.

Hirsch, C., Schajowicz, F. and Galante, J. (1967) Structural changes in the cervical spine: a study of autopsy specimens in different age groups. *Acta Orthop. Scand. Suppl.* **109**.

Jirout, J. (1973) Changes in the atlas–axis relations on lateral flexion of the head and neck. *Neuroradiology*, **6**, 215–218.

Kapandji, I.A. (1974) *The Physiology of the Joints. Vol. 3: The Trunk and the Vetrebral Column*, 2nd edn. Churchill Livingstone, Edinburgh.

Kapandji, I.A. (1982) *The Physiology of the Joints. Vol. 1: Upper Limb*, 5th edn. Churchill Livingstone, Edinburgh.

Kapandji, I.A. (1987) *The Physiology of the Joints. Vol. 2: Lower Limb*, 5th edn. Churchill Livingstone, Edinburgh.

Kauer, J.M.G. (1980) Functional anatomy of the wrist. *Clinical Orthopaedics*, **149**, 9–20.

Kelsey, J.L., Githens, P.B., White, A.A. et al. (1984) An epidemiologic study of lifting and twisting on the job and the risk for acute prolapsed lumbar intervertebral disc. *Journal of Orthopaedic Research*, **2**, 61–66.

Kessler, R.M. and Hertling, D. (1990) The ankle and hindfoot. In: *Management of Common Musculoskeletal Disorders*, 3nd edn. (ed. Hertling, D. and Kessler, R.M.). J.B. Lippincott, Philadelphia, 359–410.

Krag, M.H., Cohen, M.C., Haugh, L.D. and Pope, M.H. (1990) Body height change during upright and recumbent posture. *Spine*, **15**, 202–207.

Lambert, L.L. (1971) The weightbearing function of the fibula. A strain gauge study. *Journal of Bone Joint Surgery*, **53**, 507–513.

Larson, R.F. (1969) Forearm positioning on maximal elbow-flexor force. *Physical Therapy*, **49**, 748–756.

Lee, D.G. (1989) *The Pelvic Girdle: An Approach to the Examination and Treatment of the Lumbo-Pelvic-Hip Region*. Churchill Livingstone, Edinburgh.

Lewin, T., Moffett, B. and Vidik, A. (1961) The morphology of the lumbar synovial intervertebral joints. *Acta Morph. Neerlando-Scand.* **4**, 299–319.

Linscheid, R.L. (1986) Kinematic considerations of the wrist. *Clinical Orthopaedics*, **202**, 27.

London, J.T. (1981) Kinematics of the elbow. *Journal of Bone Joint Surgery*, **63A**, 529–535.

Lovett, R.W. (1900) The mechanics of lateral curvature of the spine. Boston M.S.J., **142**, 622–627.

Lovett, R.W. (1902) The study of the mechanics of the spine. *American Journal of Anatomy*, **2**, 457–462.

MacConaill, M.A. and Basmajian, J.V. (1977) *Muscles and Movement*. Krieger, New York.

Magoun, H.I. (1976) Osteopathy in the Cranial Field, 3rd edn. Journal Printing, Kirksville, MS.

Manter, J.J. (1941) Movements of the subtalar and transverse tarsal joints. *Anat. Rec.*, **80**, 397–410.

Marchand, F. and Ahmed, A.M. (1990) Investigation of the laminate structure of lumbar disc annulus fibrosus. *Spine*, **15**, 402–410.

McGill, S.M., Norman, R.W. and Sharratt, M.T. (1990) The effect of an abdominal belt on trunk muscle activity and intra-abdominal pressure during squat lifts. *Ergonomics*, **33**, 147–160.

Mercer, S. (1994) The menisci of the cervical synovial joints. In: *Grieve's Modern Manual Therapy: The Vertebral Column*, 2nd edn. (ed. Boyling, J.D. and Palastanga, N.). Churchill Livingstone, Edinburgh, 69–72.

Nachemson, A.L. (1960) Lumbar intradiscal pressure. *Acta. Orthop. Scand.* (supp. **43**), 1–104.

Napier, J.R. (1956) The prehensile movements of the human hand. *Journal of Bone Joint Surgery*, **38B**, 902–913.

O'Donoghue, D.H. (1950) Surgical treatment of fresh injuries to the major ligaments of the knee. *Journal of Bone Joint Surgery*, **32A**, 721.

Palmer, A.K. and Werner, F.W. (1984) Biomechanics of the distal radio-ulnar joint. *Clinical Orthopaedics*, **187**, 26.

Palastanga, N., Field, D. and Soames, R. (1994) *Anatomy and Human Movement*, 2nd edn. Butterworth Heinemann, Oxford.

Pauly, J.E., Rushing, J.L. and Scheving, L.E. (1967) An electromyographic study of some muscles crossing the elbow joint. *Anat. Rec.*, **159**, 47–54.

Pearcy, M.J. and Tibrewal, S.B. (1984) Axial rotation and lateral bending in the normal lumbar spine measured by three-dimensional radiography. *Spine*, **9**, 582–587.

Penning, L. and Wilmink, J.T. (1987) Rotation of the cervical spine. A C.T. study in normal subjects. *Spine*, **12**, 732–738.

Ramsey, P.L. and Hamilton, W. (1976) Changes in tibiotalar area of contact caused by lateral talar shift. *Journal of Bone Joint Surgery*, **58A**, 356–357.

Rockoff, S.D., Sweet, E. and Bleustein, J. (1969) The relative contribution of trabecular and cortical bone to the strength of human lumbar vertebrae. *Calcified Tissue Research*, **3**, 163–175.

Sammarco, G.J. (1989) Biomechanics of the foot. In: *Basic Biomechanics of the Musculoskeletal System*, 2nd edn. (ed. Nordin, M. and Frankel, V.H.).

Sammarco, G.J., Burstein, A.H. and Frankel, V.H. (1973) Biomechanics of the ankle: a kinematic study. *Orthop. Clin. North Am.*, **4**, 75–96.

Sammut, E.A. and Searle-Barnes, P.J. (1998) *Osteopathic Diagnosis*, Stanley Thornes, Cheltenham.

Sandzen, S.C. (1979) *Atlas of Wrist and Hand Fractures*. PSG, Littleton, MA.

Santo, E. (1935) Zur Entwicklungsgeschichte und Histologie der Zwischenscheiben in den Kleinen Gelenken. *Zeitschr. f. Anat. u Entwicklungsgesch.*, **104**, 623–634.

Sarrafian, S.K. Melamed, J.L. and Goshgarian, G.M. (1977) Study of wrist motion in flexion and extension. *Clinical Orthopaedics*, **126**, 153.

Singer, K.P. (1994) Anatomy and biomechanics of the thoracolumbar junction. In: *Grieve's Modern Manual Therapy: The Vertebral Column*, 2nd edn. (ed. Boyling, J.D. and Palastanga, N.). Churchill Livingstone, Edinburgh, 85–97.

Singer, K.P., Breidahl, P. and Day, R. (1989) Posterior element variation at the thoracolumbar transition. A morphometric study using computed tomography. *Clinical Biomechanics*, **4**, 80–86.

Snell, R.S. (1995) *Clinical Anatomy for Medical Students*, 5th edn. Little, Brown, Boston.

Steindler, A. (1955) *Kinesiology of the Human Body*. Thomas, Springfield, IL.

Stoddard, A. (1980) *Manual of Osteopathic Technique*, 3rd edn. Hutchinson, London.

Stoddard, A. (1983) *Manual of Osteopathic Practice*, 2nd edn. Hutchinson, London.

Sutherland, W.G. (1990) *Teachings in the Science of Osteopathy*. Rudra Press, Cambridge, MA.

Taleisnik, J. (1985) *The Wrist*. Churchill Livingstone, New York.

Tyrell, A.R., Reilly, T. and Troup, J.D.G. (1985) Circadian variation in stature and the effects of spinal loading. *Spine*, **10**, 161–164.

Volz, R.G., Lieb, M. and Benjamin, J. (1980) Biomechanics of the wrist. *Clinical Orthopaedics*, **149**, 112.

Weisl, H. (1955) Movements of the sacro-iliac joint. *Acta Anat.*, **23**, 80–91.

White, A.A. and Panjabi, M.M. (1978a) The basic kinematics of the human spine. A review of past and present knowledge. *Spine*, **3**, 12.

White, A.A. and Panjabi, M.M. (1978b) The clinical biomechanics of the occipitoatlantoaxoid complex. *Orthop. Clin. North Am.*, **9**, 867–878.

Williams, P.L. (ed.) (1995) *Gray's Anatomy*, 38th edn. Churchill Livingstone, Edinburgh.

Worth, D.R. (1994) Movements of the head and neck. In: *Grieve's Modern Manual Therapy: The Vertebral Column*, 2nd edn. (ed. Boyling, J.D. and Palastanga, N.) Churchill Livingstone, Edinburgh.

Wyke, B.D. (1967) The neurology of joints. *Ann. R. Coll. Surg. Eng.*, **41**, 25–50.

Yoganandan, N., Myklebust, J.B., Wilson, C.R., Cusick, J.F. and Sances, A., Jr. (1988) Functional biomechanics of the thoracolumbar vertebral cortex. *Clinical Biomechanics*, **3**, 11–18.

Zancolli, E.A., Ziadenberg, C. and Zancolli, E. (1987) Biomechanics of the trapeziometacarpal joint. *Clinical Orthopaedics*, **220**, 14–26.

Index

Page numbers appearing in italic refer to figures and tables. Suffixes in brackets, for example (-es) refer to the plural form. Unless otherwise stated, details of joint features such as bursae, fibrous capsule, fat pads and inclusions, and synovial membrane, may be found within the sub-heading 'joints' with reference to main joint areas as appropriate. Page references also include Study tasks. Assistance with terminology should be sought where necessary in the Glossary of terms and abbreviations.